EAI/Springer Innovations in Communication and Computing

Series Editor

Imrich Chlamtac, European Alliance for Innovation, Ghent, Belgium

Editor's Note

The impact of information technologies is creating a new world yet not fully understood. The extent and speed of economic, life style and social changes already perceived in everyday life is hard to estimate without understanding the technological driving forces behind it. This series presents contributed volumes featuring the latest research and development in the various information engineering technologies that play a key role in this process.

The range of topics, focusing primarily on communications and computing engineering include, but are not limited to, wireless networks; mobile communication; design and learning; gaming; interaction; e-health and pervasive healthcare; energy management; smart grids; internet of things; cognitive radio networks; computation; cloud computing; ubiquitous connectivity, and in mode general smart living, smart cities, Internet of Things and more. The series publishes a combination of expanded papers selected from hosted and sponsored European Alliance for Innovation (EAI) conferences that present cutting edge, global research as well as provide new perspectives on traditional related engineering fields. This content, complemented with open calls for contribution of book titles and individual chapters, together maintain Springer's and EAI's high standards of academic excellence. The audience for the books consists of researchers, industry professionals, advanced level students as well as practitioners in related fields of activity include information and communication specialists, security experts, economists, urban planners, doctors, and in general representatives in all those walks of life affected ad contributing to the information revolution.

Indexing: This series is indexed in Scopus, Ei Compendex, and zbMATH.

About EAI

EAI is a grassroots member organization initiated through cooperation between businesses, public, private and government organizations to address the global challenges of Europe's future competitiveness and link the European Research community with its counterparts around the globe. EAI reaches out to hundreds of thousands of individual subscribers on all continents and collaborates with an institutional member base including Fortune 500 companies, government organizations, and educational institutions, provide a free research and innovation platform.

Through its open free membership model EAI promotes a new research and innovation culture based on collaboration, connectivity and recognition of excellence by community.

More information about this series at http://www.springer.com/series/15427

P. Mary Jeyanthi • Tanupriya Choudhury
Dieu Hack-Polay • T P Singh • Sheikh Abujar
Editors

Decision Intelligence Analytics and the Implementation of Strategic Business Management

Springer

EAI
RESEARCH MEETS INNOVATION

Editors
P. Mary Jeyanthi
Bambala Institutional Area Pratap Nagar, Sanganer
Jaipuria Institute of Management Jaipur
Jaipur, Rajasthan, India

Dieu Hack-Polay
Department of Management
Crandall University
Moncton, Canada

University of Lincoln
Lincoln, UK

Sheikh Abujar
Computer Science and Engineering Department
Independent University
Dhaka, Bangladesh

Tanupriya Choudhury
School of CSE
University of Petroleum and Energy Studies
Dehradun, Uttarakhand, India

T P Singh
Department of Informatics
University of Petroleum and Energy Studies
Dehradun, Uttarakhand, India

ISSN 2522-8595 ISSN 2522-8609 (electronic)
EAI/Springer Innovations in Communication and Computing
ISBN 978-3-030-82762-5 ISBN 978-3-030-82763-2 (eBook)
https://doi.org/10.1007/978-3-030-82763-2

This Springer imprint is published by the registered company Springer Nature Switzerland AG
The registered company address is: Gewerbestrasse 11, 6330 Cham, Switzerland

Dr. P. Mary Jeyanthi would like to dedicate this book to her family, Sri. P. Prem Anand and beloved daughters Eva & Jael. Thanks to the Lord Jesus, the Almighty, for giving this opportunity and the showers of blessings.

Dr. Tanupriya Choudhury would like to dedicate this book to all Corona Warriors, for their dedication, sacrifice, and excellency towards our motherland INDIA and also he would love to dedicate this book to his parents and in-laws: Sri Mrigendra Choudhury, Smt. Minakshi Choudhury, Sri Hemabrata Bhowmick, Smt. Debjani Bhowmick and his beloved wife Rituparna Choudhury, and beloved Son Rajrup Choudhury (Ritam) for their immense support and love throughout this work. He would also like to dedicate this book to his research guides Prof. (Dr.) Vivek Kumar, Prof. (Dr.) V. Cyril Raj, Prof. Sumathy Eswaran, and Dr. Darshika Nigam, who have always mentored him during his master's and doctoral research. He would like to thank his uncle Late Girindra Mohan Choudhury for his all-time love, blessings, and support and wants to dedicate the book to his uncle. He would also like to thank his uncle(s) Dr. Tapobrata Chowdhury (MBBS) and Mr. Bhaskar Das and brothers Mr. Supriya Choudhury and Mr. Debopriya Choudhury, who supported wholeheartedly to complete this work.

Dr. Dieu Hack-Polay would like to dedicate this book to his family.

Dr. T. P. Singh would like to dedicate this book to his parents who always encouraged and supported to believe in himself, his beautiful wife Manisha, a person who is always behind him whenever he needs, and his son Anurav, who is the motivation of his life.

Mr. Sheikh Abujar would like to dedicate this book to his family that includes his lovely mother, Farida Momin; Father, Sheikh Abdul Momin; his most beautiful wife, Parsha Sanjana; two caring sisters, Nadia and Nusrat; and talented niece, Raisa, and nephew, Jabir. They all are the precious blessing from Allah and everything of his life. He also wants to dedicate to his teacher, Prof. Syed Akhter Hossain, with love. "Thank you, Sir, for being a great teacher in my life."

Foreword

This book is the result of a framework for developing an analytics strategy that includes everything from problem definition and data collection to analysis of business and decision-making. This edited volume focuses on faculties and managers who want to become experts of analytics in Decision Intelligence and provides strategic management support. In India, this is the first book on Decision Intelligence to implement strategic and business management with real-time case studies. The idea of this book is to learn and implement. The chapters comprise techniques and methods of Decision Intelligence analytics with Business Intelligence and Artificial Intelligence in right decision-making to implement Strategic Management and cases, and discussions with real-time scenarios. All the best and congratulations to the entire editorial team and waiting for the next volume too. We truly believe that the book will fit as a good read for those looking forward to exploring areas of decision intelligence.

Associate Professor, Amity University Tashkent, Uzbekistan

Praveen Kumar

Preface

Technological developments, with increasing automation, have gathered pace in the past few decades and have significantly accelerated in the last 20 years or so. These have transformed daily lives and organizations and the world of work more broadly. Thus, technology is pervasive in society and industry, aiding the way in which work tasks are approached and the speed of execution.

In recent years, the advent of artificial intelligence (AI) has been an added dimension of technology, which not only facilitates work tasks, but more significantly aids human intelligence and decision-making. More and more organizations are drawing on the benefits of AI which is revolutionizing business strategic formulation and decision-making. It is thought that artificial intelligence helps make business decisions more quickly but is also qualitatively an advantage. It helps to deal with complexity in a way that its creator, the human being, cannot easily do. Examples of such complexity include scheduling, customer relations management (CRM), Intelligent Cloud Services, Blockchains, etc.

In practical terms, manufacturing has witnessed the introduction of a package like Industry 4.0 which has been credited with increasing quality of production, level of productivity, and flexibility in the sector. In the medical sector, with the advent of coronavirus disease (COVID-19) in 2020, a great realization of the weight of intelligent technology has been noticed. For example, the use of ventilators, artificial intelligence, the Cloud, and Big Data to monitor and forecast patients' conditions has shown that intelligent technology is inextricably linked to efficiency in human organizations in the twenty-first century. Another example is the Automatic Computing environment which helps to deal with threats and critical problems effectively. Because the Automatic Computing environment is self-regulating and self-managing, it represents a major resource capability to human organizations. Thus, for most organizations, automation is increasingly becoming one of the sine qua non conditions for competitive advantage. The examples of the criticality of automation are numerous and stretch across sectors of activities.

This book is a significant addition to the body of knowledge emerging to profile the novel reality of artificial intelligence and decision analytics. The various chapters that it contains attempt to each deal with a specific area of organization where automation aids creativity and efficiency in decision-making. This deliberate choice of a diversity of fields of coverage is to emphasize the application of decision intelligence to almost all aspects of contemporary society and assist those working in the sectors in understanding the working and strategic opportunities afforded by artificial intelligence in their domain of activity.

We hope that readers will draw many benefits from this book's theoretical and practical aspects to enhance their own practice or research. The book's coverage is as follows:

Chapter 1 discusses the available analytics in the context of Decision Intelligence Analytics with the implementation of Strategic Business Management. **Chapter** 2 discusses how organizations can adopt analytics techniques in order to completely change their operations and strategies to match up with the era where data plays a huge role in making informed decisions. **Chapter** 3 delineates the basics of AI with its solicitation in healthcare, education, agriculture, transportation, smart city projects, manufacturing industries, and retail. It also evaluates the challenges of AI's interaction with management and organizations for decision-making. **Chapter** 4 focuses on the role of artificial intelligence in strategic formulation, SWOT analysis, and use of AI in strategic management activities. **Chapter** 5 explores how workforce planning, safety, and recruiting could use technology effectively and provides some guidance for the application of decision intelligence analytics in the implementation of strategic business management. **Chapter** 6 includes practical and theoretical perspectives on AI and provides room to recognize the state of the art and the main developments and trends in the industry and services. The chapter also provides insights into AI policies. **Chapter** 7 explains how decision intelligence can be a framework that brings advanced analytics and machine learning techniques to the desktop of the nonexpert decision-maker to incorporate in data science to overcome dynamic business problems and articulated in black swan theory. **Chapter** 8 explains how analytics assists in anticipating numerous prospects and permits organizations to survey various potential results dependent on their activities. **Chapter** 9 presents panel data in terms of a lattice and explains various rule-based analysis with support and confidence level. The chapter also exemplifies how algorithm is implemented in a data set and various interesting observations are generated. **Chapter** 10 focuses on the emergence of business intelligence and its role. The chapter provides the necessary guide for the decision-makers to be in a state of readiness to set up strategies or modify them based on real-time data. **Chapter** 11 deals with the way in which DI and strategic business planning (SBP) can place the organization in a dynamic competitive seat. **Chapter** 12 provides an understanding of how social media platforms play an important role in providing real-time streaming data for business analysis. The chapter aims to provide a pathway to undertake social media analytics. **Chapter** 13 is an earnest attempt to explore the people-analytics interface and the journey from predictive analytics to prescriptive analytics. **Chapter** 14 explains how predictive analytics with the usage of machine learning algorithms is helping different business sectors to make informed and better decision based on past and current records. **Chapter** 15 discusses issues concerning virtual personal assistants, email spam, online fraud detection, traffic predictions, social media personalization, and many more. **Chapter** 16 focuses on the significance of machine learning techniques in the recognition and reporting of behavioral biases (cognitive) among traders. The chapter outlines the vital role played by research in analytics and psychology in helping traders to improve their financial performance and avoid suboptimal decisions. **Chapter** 17 focuses how applying business intelligence techniques and approaches produces enhanced performance that helps business owners to understand the trends and find the loopholes of the sales process. **Chapter** 18 is written with the investors in mind to assist them in making choices about their portfolios of investment into the Japanese Yen forex market using decision intelligence. **Chapter** 19 examines the language impairment problem regarding the Broca's area of the brain by focusing on an analysis of the behavior of autism and discusses the role of current technologies in treatments.

The volume is edited by an international team of academics and experts from various countries, emphasizing the global appeal of the coverage. The team comprises:

Jaipur, Rajasthan, India — P. Mary Jeyanthi

Dehradun, Uttarakhand, India — Tanupriya Choudhury

Moncton, Canada
Lincoln, UK — Dieu Hack-Polay

Dehradun, Uttarakhand, India — T P Singh

Dhaka, Bangladesh — Sheikh Abujar

Acknowledgments

Dr. P. Mary Jeyanthi would like to thank the Lord Jesus, the Almighty, for His showers of blessings throughout the research work to complete the book successfully. She is highly grateful to all the authors from all parts of the world for contributing their research work for this book. She would like to say thanks to her husband P. Prem Anand for his constant encouragement. She would also like to thank all the staff of the Research section of Jaipuria Institute of Management Jaipur for their kindness. She thanks the management of the Institute of Management Technology, Nagpur, for their support to do this work. Finally, she would like to acknowledge all those who contributed directly or indirectly to complete this book.

Dr. Tanupriya Choudhury and Dr. T. P. Singh would like to thank their workplace, University of Petroleum and Energy Studies, Dehradun, India, for giving a positive research environment to start this proposal. They would like to thank all contributors from twelve different countries and specially reviewers throughout the globe who helped to review the chapters to maintain the quality of the book and for their valuable suggestions always whenever required. They are also thankful to the senior leadership of the University of Petroleum and Energy Studies (UPES) and administration for giving the opportunity to hold DIASBM2020 and providing with all possible support. They are thankful to Shri Sharad Mehra, CEO, GUS-Asia, for his "all-time-go-ahead" blessings and freedom of work. Shri Dr. S. J. Chopra, Chancellor UPES for his blessings and guidance as always. Honorable Vice-Chancellor Dr. Sunil Rai has been a continuous support as a torchbearer to us, big thanks to you Sir for your mentorship. They would also thank their colleagues and friends for all-time support, specially Dr. A. Mansurali from PSG Institute of Management, Coimbatore, Tamil Nadu, India, Mr. Partha Chakraborty (HoD CSE) from Comilla University Bangladesh, Dr. Durga Prasad Sharma, Dr. Bhupesh Kumar Singh from AMU Ethiopia, and Dr. Praveen Kumar from Amity University Tashkent for moral and technical support as always whenever required. They would like to thank everyone enough for their involvement and their willingness to take on the completion of tasks beyond their comfort zones. See you all in the next edition of the book.

Dr. Dieu Hack-Polay would like to thank his family for all-time support. He is grateful to his workplace, University of Lincoln (UK) & Crandall University (Canada), for supporting and providing him with such a working environment that he could devote time for such projects. He would like to thank everyone enough for their involvement and their willingness to take on the completion of tasks beyond their comfort zones. See you all in the next edition of the book.

From Mr. Sheikh Abujar's point of view, this book has been truly a collaborative effort, with different scholars and motivated by the aspiration to highlight the advanced AI theories and their applications. He would like to thank Independent University, Bangladesh (IUB), Daffodil International University (DIU), DIU NLP and ML research LAB and Computational Intelligence LAB (CiLAB) for all their support.

From all Editors,
The team of Editor(s) would like to express their sincere thanks to the entire publishing team from Springer and EAI for their wonderful assistance always during the long journey of publishing this book at every step. Special thanks go to Eliška Vlčková (Managing Editor from EAI), Mary E. James (Senior Editor, Applied Sciences, Springer), Brian Halm (Project Manager, Springer Science and Business), Pradheepa Vijay (Springer Team), and Savita Rockey Samuel (production in charge from Straive) for their herculean task of editing and numerous follow ups which happened during this task. We all feel highly motivated to work with you and look forward to accomplishing similar projects in future too.

Contents

About the Editors

P. Mary Jeyanthi is an Associate Professor of Jaipuria Institute of Management, Jaipur, India. She had one decade of industry experience in HDFC Bank Ltd, Business Intelligence Unit, Chennai, Tamil Nadu. She had a handful of expertise in business intelligence and forecasting analytics. She had done her research in "A New Implementation of Mathematical Models with Metaheuristic Algorithms for Business Intelligence." She is responsible for Business Intelligence and Analytics portfolio creation and execution of different managerial strategies. She is also a visiting faculty of IMT Ghaziabad and Woxsen University, Hyderabad. Besides that, she had published a number of research papers in various reputed national and international journals. She conducted several workshops on research methodology and business intelligence. She has more than 10 years of experience in information technology in the banking sector with extensive skills and experience in business intelligence. Being an entrepreneur, she has intensive knowledge in the research field. She served as the resource person in many colleges conducting soft skills training and workshops related to research paper writing. She had contributed many articles to magazines related to the building up of human capital and business analytics. She has expertise in the novelty approach of writing and managing the analytics with real-time scenarios. Her research interests include business intelligence and analytics, artificial intelligence, big data analytics, management information systems, data analysis, and business models.

Tanupriya Choudhury received his bachelor's degree in CSE from West Bengal University of Technology, Kolkata, India, and master's degree in CSE from Dr. M.G.R. University, Chennai, India. He has received his PhD degree in the year 2016. He has twelve years of experience in teaching as well as in research. Currently, he is working as an Associate Professor in the Department of CSE at UPES Dehradun. In 2018, he has received the Global Outreach Education Award for Excellence in best Young researcher Award in GOECA 2018. His areas of interests include human computing, soft computing, cloud computing, data mining, etc. He has filed 14 patents to date and received 16 copyrights from MHRD for his own software. He has been associated with many conferences in India and abroad. He has authored more than 85 research papers to date. He has delivered invited talk and guest lecture in Jamia Millia Islamia University, Maharaja Agrasen College of Delhi University, Duy Tan University, Vietnam, etc. He has been associated with many conferences throughout India as TPC member and session chair, etc. He is a lifetime member of IETA, member of IEEE, and member of IET(UK) and other renowned technical societies. His association with the corporate sector includes Technical Adviser of DeetyaSoft Pvt. Ltd. Noida, IVRGURU, Mydigital360, etc. He holds the position of Secretary in IETA (Indian Engineering Teacher's Association—India) and the Adviser Position in INDO-UK Confederation of Science, Technology and Research Ltd., London, UK, and International Association of Professional and Fellow Engineers, Delaware, USA.

Dieu Hack-Polay is an Associate Professor of Management at Crandall University (Canada) and an Associate Professor in Organizational Studies at Lincoln International Business School, University of Lincoln (UK). He has significant industry experience. His academic experience spans several years in various institutions and countries. Dr. Dieu Hack-Polay obtained a Doctor of Philosophy (PhD) in sociology at the University of Surrey and a Doctor of Education (EdD) in leadership and management at the University of Lincoln. His main research interest is in sociology and international human resource management. He has authored three books and several book chapters as well as research articles in international journals. His latest publications examine the role of technology in environmental sustainability. His professional affiliations include chartered fellowship of the Chartered Institute of Personnel & Development (CIPD), fellowship of the Higher Education Academy (HEA), and fellowship of the Society for Education & Training (SET). He is also an active member of the British Sociological Association (BSA) and the Canadian Sociological Association (CSA).

Thipendra Pal Singh is currently positioned as Professor and Head of the Department at the School of Computer Science, University of Petroleum & Energy Studies, Dehradun, UK. He holds a Doctorate in Computer Science from Jamia Millia Islamia University, New Delhi. He carries 22 years of rich experience with him. He has been associated with Tata Group and Sharda University, Greater Noida, NCR, India. His research interests include machine intelligence, pattern recognition, and development of hybrid intelligent systems. He has guided 15 master's theses. Currently, 05 research scholars are working towards their doctoral degree under him. There are more than dozens of publications to his credit in various national and international journals. Dr. Singh is a member of various professional bodies including IEEE, ISTE, IAENG, etc. and serves on the editorial/reviewer panel of different journals.

Sheikh Abujar is a Lecturer in the Department of Computer Science and Engineering, Independent University, Bangladesh (IUB). Broadly, his methodological research focuses on the application of machine learning and data mining. Sheikh Abujar currently works on different areas of natural language processing and adaptive algorithms. He has published his research works in several Scopus Indexed proceedings of IEEE, Springer, and Elsevier, and he has served as a reviewer for several international conferences and journals. Many of his and his students' research work have won "Best Research Paper—Award" in many international conferences. Prior to joining IUB, he was a faculty member at Daffodil International University (from September 2017 to January 2021) and at Britannia University (from February 2015 to August 2017). During his career at DIU, he was Founder and Supervisor of the DIU-NLP and Machine Learning Research LAB.

Chapter 1
Analytics Techniques: Descriptive Analytics, Predictive Analytics, and Prescriptive Analytics

Ashish K. Sharma, Durgesh M. Sharma, Neha Purohit, Saroja Kumar Rout, and Sangita A. Sharma

1.1 Introduction

A business is an enterprise-oriented entity where services and products deliver to its customers, and customers are either paid in the form of money or exchanged with other services and products. The main principle of business is to cover all stakeholders like customers, its employees, of course owners, and even directly and indirectly must cover society as a whole too. Ultimately, the main goal of any business is to get maximum benefits, fulfill stakeholders' needs with less input, and gain competitive advantage over others. This calls for strategic business management. A business with short-term objectives and goals will fight to set corporation direction, focus efforts, and acquire competitive benefit. Yet by applying strategic management, organizations can not only live, but flourish. Here is why strategic management can improve performance. Strategic management is the management of a business's resources to effectively accomplish its objectives and targets. It is a proposal to act to ensure performance goals are achieved, and the business remains

A. K. Sharma (✉) · N. Purohit
Department of CSE, G H Raisoni College of Engineering (GHRCE), Nagpur, India
e-mail: ashishk.sharma@raisoni.net; neha.purohit@raisoni.net

D. M. Sharma
Department of AI, G H Raisoni College of Engineering (GHRCE), Nagpur, India
e-mail: durgesh.sharma@raisoni.net

S. K. Rout
Department of CSE, Gandhi Institute For Technology, Bhubaneswar, India
e-mail: rout_sarojkumar@yahoo.co.in

S. A. Sharma
Info Origin Technologies Pvt. Ltd, Gondia, India
e-mail: sangitaasharma1975@gmail.com

P. M. Jeyanthi et al. (eds.), *Decision Intelligence Analytics and the Implementation of Strategic Business Management*, EAI/Springer Innovations in Communication and Computing, https://doi.org/10.1007/978-3-030-82763-2_1

to grow. Strategic management suggests complete path by developing strategies and rules designed to attain targets and then assigning means to implement the tactics. Ultimately, strategic management is for organizations to increase a competitive edge over their rivals [1]. Strategic business management focuses on setting goals, surveying the competitive surroundings, organization internal operation, assessing strategic planning, and making sure that management spreads out the strategic planning across the organization through wise decision-making processes [2].

Decision-making drives the businesses and ensures better performance. The decision-making process designs the roadmap to reach its customer desires and indirectly reaches its business goal. Certainly, a good decision brings organization, administrators, and branches nearer to their target and even resolve all involved issues too. The decision-making forms the basis in any businesses. The significance of decision-making for successful businesses in ecommerce, specifically for online shopping websites design and forecasting of customer requirements has been highlighted by [3–5]. Decision intelligence is a new buzzword that acts as a better framework for decision-making models. It embeds the machine learning methods into decision-making processes to serve the purpose. It comprises decision-making constituents that brings to concrete decision with the help of executing, modeling, designing, adjusting, and tracking the processes [6].

To make business decisions, the decision intelligence has functionalized artificial intelligence and machine learning for better understanding the unpredictable failures. By bringing computer science, big data, psychology, neural science into a single platform, it can transform the meaning of decision-making completely in politics, business, government agencies, and defense or even where humans think and make decisions [7].

Organizations often have to make firm decisions for better efficiency, threat management, and profits. Premium quality, quick, and prompt decision-making process strengthen the organization efficiency and performance, and proper threat management is required failing which it may lose the organization projected objectives such as following the Regulatory Acts, Economics, etc. In the same way, a good decision maker system assists to flourish and increase its profits as well, and that is why an organization takes firm and professional approach for decisions like SWOT analysis (Strength, Weaknesses, Opportunities and Threats) processes that lead to business success. It has been done for over 50 years for managing threats, planning, and decision-making as well [8]. It identifies the internal and external threats and opportunities in its business world. In addition, the amalgamation of a growing complex business world and the desire to spread data extensively and to gain competitive edge over competitors have now led organizations to concentrate on applying analytics to operate strategic business decisions successfully. In present business scenario, each organization is in quest of a way to have their decision-making more effective and business analytics offers them this benefit.

Analytics [9] is the computation-based analysis method that interprets the old data by applying the specific method and discovers the meaningful information that makes good predictions, understanding the business strategy, and even improves the existing methodology too. Analytics [10] can assist businesses not only in making better strategic decisions but also in attaining good functional efficiency, higher customer satisfaction, and robust profit and revenue levels. For sustaining its business in a competitive environment, proper analytics [11] strategies are required that give numerous advantages that become more eco-friendlier by using efficient technology oriented resources, by offering quality products and services and of course sustainability oriented methodology [12]. However, with various analytics available in the market like descriptive, predictive, and prescriptive, each having the ability to use the data to answer different questions, it is necessary to explore these analytics and their applicability in context with the strategic business management. To this end, this chapter discusses the available analytics in context with the decision intelligence analytics with the implementation of strategic business management.

1.2 Analytics

Earlier, the processing speed and data stowage have confined analytics. However, now those restrictions no longer apply, generating the way to more complicated deep learning and machine learning algorithms that can control huge amounts of data in several phases. Thus, the typical descriptive, prescriptive, and predictive capabilities of analytics have seen considerable rise with self-learning and automatic technology, steering in the artificial intelligence age. Analytics is essential to numerous functional parts and skills and is now treated as a strategic asset by most companies. It employs calculation and data to reply business queries, determine relationships, forecast unknown results, and automatize the decisions. In addition, it can be utilized to discover meaningful behavior in the data and expose novel information depending on machine learning methods, statistics, predictive modeling, and functional mathematics [13]. Analytics is the planned and proper computational-based analysis using the data and statistics. It helps an organization to understand the effects of changing the marketing methodology which is invaluable while enhancing the business, detection of risk, healthcare, digital advertisement, or many more areas where it can be applied. Analytics has prominent software ACL, IDEA, SPSS, Minitab, R, R+, SAS, IBM Watson, etc. All these software have statistical capabilities. However, using these softwares requires one to have a very good understanding of statistics.

1.2.1 Role of Analytics in Business

The process of analyzing the business refers to business analytics or analytics in the company [14]. Analytics helps businesses make data-driven decisions. It lets companies to automate their complete decision-making process, ensuring delivery of real-time responses whenever needed. Moreover, it aids to gain crucial business insights by offering the right data to work it. It brings a revolution in the technology, competencies, and applications to investigate data constantly, and as a result, this drives the business decision by understanding competently [15]. The technology-based approach keeps the company close to its customers. All the above types of analytics outcome help to discover possible opportunities in the targeted business [16].

1.2.2 Types of Analytics

Businesses employ analytics to study, assess their data and transform its conclusion into insights that finally assist managers, administrators, functional employees, and executives that improve business decisions. There are various analytics techniques that businesses use in strategic business intelligence such as:

1. Descriptive analytics
2. Predictive analytics
3. Prescriptive analytics

1.3 Descriptive Analytics

Descriptive analytics is a kind of analytics that delivers the "What happened?" information regarding an organizations processes. It works on the past data to identify the changes that have happened in the business. It explains the use of a variety of historic data that helps businesses to draw comparisons. If economic metrics are to be considered, then most usually stated monetary metrics occurs to be product of descriptive analytics—like, year-over-year change in prices, month-over-month sales progress, the number of consumers, or each subscriber's total income. All these are measures that let businesses know what has happened in a business in a fixed period. It analyzes the raw data to make meaningful conclusions which are valuable for various stakeholders like investors, managers, customers, staffs, etc. It utilizes the wide range of data to get clear idea of what has occurred, how it varies with other. With the help of this wide range of old data, it helps to acquire a comprehensive review on line of action and efficiency that can be based on business strategy [17].

Aggregation and mining the data are important steps used in descriptive analytics. Data aggregation is done first to gather and sort the data so that the analyst can manage the datasets efficiently. Later, the data mining is used to extract meaningful data. It identifies and understands the patterns that have been discovered by applying some intelligent methods. It becomes essential to represent the data visually, after transforming, sorting, and analyzing it. Therefore, descriptive analytics is used to set the sensitive measures along with business targets to evaluate the business' present conditions based on activities that occurred.

Following steps are to be executed to make a successful project based on descriptive analytics:

1. Decide the business benchmark to assess the analytics based system performance.
2. The necessary data is recognized.
3. The data are gathered and put in order to process.
4. The data is examined to discover patterns and calculate its efficiency.
5. The data is visualized by presenting the discovered pattern in the form of graphs and charts that naive user can understand easily.

Today the businesses have gone more data driven. These data come from people who share their views, expressions, experiences, emotions that become possible for the firms to understand customers' views. Decision-making is a vital process when it comes to reach out a final decision from the number of alternative ways [18]. Decision-making has become a data-driven process, and this data makes statistics if some specific method has been applied to reach a proper decision, and to reach a good decision requires a large dataset that can make a proper result. A large dataset is not the only requirement for the decision-making even though it requires a proper analytics method that can retrieve a proper decision smartly and intelligently. Two different traditional decision-making techniques were proposed, viz. programmed and non-programmed decisions [19].

1.3.1 Functions of Descriptive Analytics

Descriptive analytics comprises diverse statistical functions, like suppression summary statistics, and regression analysis. Following are the functions that make descriptive analytics operational [20]:

1. *Business Metrics and KPIs*: This includes identification of the key performance indicators (KPIs) that is to be measured to attain business targets like reducing prices, increase income or better understanding production. Thus, a KPI to calculate income would be items sold.
2. *Data Gathering and Aggregation*: After identifying the business targets and relevant KPIs, then the next phase is to determine the data sources for such information as the data held by businesses are usually in many locations, like databases, desktops, and more. This generates need for organizations to catalog data.
3. *Data Extraction*: Data extraction is a tedious task. It includes data transformation, data cleaning, data replication, and more. To carry out these tasks, the data automation tool is usually used.
4. *Data Analysis*: After the data is organized properly, it needs to be analyzed. To offer insight, the data analysis generally connects the number with business metrics.
5. *Data Presentation*: Once the picture is painted, it should be shared with stakeholders to inform decisions. Data presentation is carried out by data visualization and presentation, via charts and graphs.

1.3.2 Advantages of Descriptive Analytics

Through statistics and summarization, descriptive analytics provides the ability to assess things in a more healthier way as to how processes are functioning, to check whether business targets are being achieved most proficiently. This helps in business perform better. Some of the benefits offered by this analytics are as follows:

1. *Delivers Historical Context*: Through descriptive analytics, businesses can analyze the past data to gain insight on how consumers and products relate to one another. It results in predictive analytics that directs companies in moving forward.
2. *Measure Business Goals*: Through KPIs, the descriptive analytics can demonstrate how present processes are functioning. This helps in assessing the business goals.
3. *Holistic Approach*: Descriptive analytics aids to recognize trends and then visualize patterns. This helps to classify an organization's strengths and weaknesses and thus offer a historical summary to support function more optimally in the future.

1.3.3 Descriptive Analytics and Its Uses

To have more clear view on descriptive analytics, let us see an example of an online review of any product, the descriptive analytics could be applied to evaluate the number of consumers involved in the review forum. In addition, it can be used to know how many times the individual consumer shared in the forum. Descriptive analytics can be employed to better understand the customers' behavior by segmenting the customers into dissimilar audiences and then tailoring the marketing strategies. Some other examples of how descriptive analytics can be used include the following: summarizing past events such as sales and operations data or marketing campaigns. Another example as cited by ScienceSoft's practice says that if total number of metal parts manufactured in each month, income on each group of product, and monthly profits are analyzed, then it would help manufacturer to respond to a sequence of "what happened" queries, helping them focused on product classes. This analytics can help bank in assessing and revealing the loan risk with their customers. Like, when the loan rate of interest is high, the married struggles more than the single one and hard to repay it. Hence, if the rate of interest increases in the future, then the bank may be able to reduce the probable risk. Thus, the bank can use descriptive analytics to better recognize its exposure to risk.

1.3.4 *Need for Other Analytics*

Descriptive analytics parses historical data to better recognize the changes that have happened in a business. Using a variety of historic data and benchmarking, decision makers get a complete view of performance and trends on which to base business strategy. Thus, businesses can utilize the descriptive analytics for more improved decision-making leading to successful business. Descriptive analytics manipulates raw data from several sources of the data to provide actionable insights into the past. But, these findings just specify that something is incorrect or correct, without focusing on why. As a result, only descriptive analytics is often not recommended to highly data-driven companies. Rather they are suggested to be used with other data analytics for better outcome [21].

1.4 Predictive Analytics

Predictive analytics is about using past data; ML and AI predict what will happen in the future. The objective is to give a finest evaluation of what will happen in near future. The past data is given to a mathematical model that takes into account key patterns and trends in the data followed by application of model to present data in order to forecast what will occur next [22]. This analytics can be employed to decrease risks, enhance operations, and increase revenue. Thus, more organizations are now inclined to predictive analytics as it can resolve complex problems and reveal new opportunities [23]. It is progressively being used to model everything. Consumer behavior patterns, weather forecasting, predicting political events, predicting the course of diseases in patients to name a few. It is based on data and facts as well as information indicating that it is scientific and methodical and not based on instinct and rumor. This sets it different from other methods [24].

Predictive analytics has bagged the support of various organizations, with a global market projected to achieve around $10.95 billion by 2022 as per 2017 report by Zion Market Research. It helps organizations to sift through current and past data to discover trends, predict events and situations that should happen at a particular time, based on provided factors. Moreover, organizations can detect risks and opportunities through finding and exploiting patterns present within the data. Models can be drawn, to learn associations between several behavior features. Such models allow the evaluation of promise or threat presented by a particular set of situations, managing informed decision-making across several classes of procurement events and supply chain [25]. Predictive analytic models support rational decision-making, limiting the threat of biases in decisions [26]. Asniar et al. proposed predictive analytics to forecast consumer behavior by using behavior informatics and analytics method in order to improve business decision making [27].

1.4.1 Steps in Predictive Analytics

According to [28], following are the steps involved in the Predictive Analytics:

1. Recognize what you want based on historical data.
2. Next, check availability of data with you to answer those questions.
3. Train the module to learn from our data and forecast the results.
4. Schedule modules.
5. Use the forecasts and perceptions to move on these decisions.

Predictive analytics can prove beneficial to the companies in targeting customers on the basis of their behavior. Companies may use this analytics to collect data on consumers followed by predicting future actions based on past behavior. This information obtained can then be used to make decisions that influence the business's bottom line and impact outcomes.

1.4.2 Predictive Analytics and Its Uses

According to [29], following are some common uses of predictive analytics:

- *Enhancing marketing movements* to determine buyer reviews to marketing movements or buying behavior.
- *Making better processes* to efficiently manage accounts and other means, or to fix costs for services.
- *Monitoring fraud*. Analytics can detect movement and catch or note uncommon or out of the usual buyer movement, often immediately.
- *Mitigate risk*. Today, suppliers like car merchants, use beyond the credit score to decide whether to finance someone. Likewise, they are used to things such as driving records and insurance claims to elect if the customer is a threat.

1.4.3 Predictive Analytics Examples

Predictive analytics has been widely used by organizations today in a number of ways. It is now being used in fields such as healthcare, retailing, finance, hospitality, etc. [25]. Predictive analytics can also be used to reduce crime, combat terrorism, and solve trivial healthcare issues related to how diseases can spread in some places, and the prediction for their spread to other areas and patients [28] has cited some real-life examples depicting how predictive analytics can be better utilized. These are as listed below:

- Identifying customers that may stop a service or product.
- Refer marketing movements to buyers who are inclined to purchase.

- Improve customer service through proper planning.
- Aiding sellers' goal as to which consumers will cancel the buying or indulge in the purchasing process.

The important point about predictive analytics is that it is about knowing the unknowable which means that the known unknowns can be transformed into known knowns. This supremacy qualifies it to be a game changer in the field of business. Predictive analytics can strongly assist to transform your company by helping you achieve key strategic objectives by competing more efficiently to protect strongly and distinctive competitive stronghold through increasing sales revenues, retaining customers, and determining new buyer bases. It may include applying security strategies by better avoiding fraud, noticing, and managing and refining processes and advancing our core business volume to become more competitive. It paves way for learning key insights that can improve in making business policies. It focuses on to satisfy customer expectations and acting on insights to attain anticipated outcomes [30]. It utilizes all kinds of data to reach at models that envisage human behavior. It can be extended to explore what inspires customers to buy or not buy products as well as map evolving trends in marketing and business strategy [31].

1.5 Prescriptive Analytics

Prescriptive analytics emphasizes on discovering the best way given the available figures. It relates to both predictive and descriptive analytics, but it stresses on actionable insights in place of data observing. Descriptive analytics provides BI insights into what has occurred, and predictive analytics focuses on predicting possible results, and prescriptive analytics targets to discover the best result among a variety of options. Moreover, the field also empowers companies to make decisions based on optimizing the outcome of future events or threats, and offers a prototype to learn them. It is basically a statistical technique used to draw decisions and produce recommendations based on the computational results of algorithmic prototypes [32].

Prescriptive analytics analyzes the raw data to make good decisions in businesses. Especially, it factors information about probable scenarios or circumstances, previous performance, present performance, and available resource, and suggests strategy or plan. This analytics can be used to make decisions on any time horizon. Prescriptive analytics is not restricted to predicting future outcomes only. Rather, it goes beyond it and suggests actions to benefit from the predictions depicting the decision makers the consequences of each decision choice. It not only predicts why it will occur and what will occur but also when it will occur [33]. Prescriptive analytics is evaluated as the top level of data analytics. It uses optimization to recognize the best method to reduce or increase the intended objective. It needs a predictive model with actionable data and a feedback system that tracks the results generated by actions [26]. The prime objective of prescriptive analytics is to suggest what act to be done to address the upcoming problems. It is carried out after predictive analytics to aid business and understand the fundamental reasons of difficulties and plan and suggest the best way. It discusses insights on possible outcomes and results that in turn maximize the main business metrics. It works by merging numerous business rules, data, and mathematical models. It can have a major impact on the overall processes and upcoming business development. Organization uses this to fix charges for the products, create plans, and determine the locations for bank branches.

Prescriptive analytics can prove beneficial to healthcare strategic planning by analytics to leverage the operational and usage data combined with external aspects like people demographic trends, financial data, and people health trends, to more precisely plan for future capital investments like new facilities and equipment utilization. It helps to understand the trade-offs between adding extra beds and increasing an existing facility or going for constructing a new one. To design and implement an effective prescriptive analytics strategy, an organization needs an information management strategy that includes both internal and external data as well as both structured and unstructured data, a technology strategy, and a data science strategy [34].

1.5.1 Advantages of Prescriptive Analytics

Prescriptive analytics can assist to increase efficiency, avoid fraud, meet business goals, limit threat, and make more reliable consumers. However, it is not perfect. It proves effective only if companies know what questions to ask and how to reply. It depends largely on the input assumptions and thus if they are not proper or valid, the results also will not be correct. When used effectively, however, it can help companies make decisions based on examined facts. It can simulate the likelihood of various outcomes and show the probability of each, helping organizations to gauge the level of risk and uncertainty they face than they could be relying on averages. Organizations get to know the likelihood of worst-case scenarios and plan accordingly [33]. In addition, prescriptive approach offers some benefits such as:

1. Long-term/strategic thinking
2. Coordinating effort
3. Learning from each other

1.5.2 Prescriptive Analytics and Its Uses

Many data-intensive government agencies and businesses can benefit from using prescriptive analytics. This includes healthcare and financial services domains where the cost of human error is high. It can be a great tool for fire department as it could be employed to evaluate to know if a local fire department needs residents to empty a particular area when a wildfire is burning nearby. It could also be used to forecast chances of popularity of article based on data about searches and social shares for related topics. Another use could be to regulate a worker training program in real time based on how the employee is replying to each lesson [33].

1.5.3 Prescriptive Analytics Examples

The best example of this analytics is in a traffic domain that enables one to choose the best path to home, considering the distance of the shortest path, the speed of traveling, and prevailing traffic conditions in the town [35]. Another example is energy and utilities. Natural gas rates vary dramatically depending upon geo-politics, demand and supply, econometrics, and climate conditions. Transmission (pipeline) companies, gas producers, and utility firms are eager to forecast the precise gas rates so that they can lock in favorable terms while avoiding risk [34]. Google's self-driving car, Waymo, is an example of prescriptive analytics in action. On each tour, the car does number of calculations that assists the car decide when to change paths, whether to slow down or speed up, and when and where to turn—the same decisions a human driver makes behind the wheel [36]. Additionally, the prescriptive analytics can be applied in the following sectors for effective decision-making and more improved outcomes [33, 37].

- Hospitals and clinics
- Airlines
- Sales
- Higher education
- Banking
- Retail

1.6 Conclusion

Strategy business management can be achieved through effective decision-making.

Of late, the organizations often have to make firm decisions for better efficiency, threat management, and profits. In addition, the amalgamation of a growing complex world and the desire to spread data extensively and to gain competitive edge over competitors has now led organizations to focus on applying analytics so as to operate strategic business decisions effectively. Decision intelligence provides a framework for best practices in organizational decision-making.

In light of the above, this chapter has discussed the available analytics in context with the decision intelligence analytics with the implementation of strategic business management. Descriptive, predictive, and prescriptive analytics have been discussed in this chapter. The study highlighted the applicability and areas of each of this analytics and their ability to use the data to answer different questions. It is realized that the incorporation of decision intelligence analytics with strategic business management decreases the functional risk and helps to forecast the profits and revenues. The study reveals that all these analytics can aptly guide the organizations in efficient decision-making which in turn will help in delivering strategic business management. It is realized that the chapter will assist the user to gain insights on various analytics, their scope, and applicability and how they can be applied for more improved decision-making leading to successful business.

1.7 Future Directions

Integrating decision intelligence analytics with strategic business management reduces the functional risk and helps to forecast the profits and revenues. However, this giant integration can be utilized in the future for the following:

1. Artificial intelligence-based system might be prescribed the perfect solution, can be implemented crucial decisions, even it will suggest to modify the existing Optimization Algorithm and subsequently include all those knowledge will recognize all expected and unexpected situations and try to take out from them. Eventually, through this integration, the future system will become self-dependent, self-educated, self-instructor.
2. Through this integration, opportunities and threats will be understood in advance, it might also be possible to think beyond the organization walls itself and will help to design the complete business model.

References

1. Victoria University's Master of Business Administration. What is strategic management and why is it important? https://online.vu.edu.au/blog/why-strategic-management-important. Accessed 16 Dec 2020
2. W. Kenton, Strategic management, 2020. https://www.investopedia.com/terms/s/strategic-management.asp. Accessed 16 Dec 2020
3. A.K. Sharma, I.C. Mehta, J.R. Sharma, Development of fuzzy integrated quality function deployment software—a conceptual analysis. Manager's J. Softw. Eng. **3**(3), 16–24 (2009)
4. S.K. Purohit, A.K. Sharma, Database design for data mining driven forecasting software tool for quality function deployment. Int. J. Inform. Electr. Business **7**(4), 39–50 (2015)
5. A.K. Sharma, S.P. Khandait, A novel fuzzy integrated customer needs prioritization software tool for effective Design of Online Shopping Websites. International journal of operations research and information systems (IJORIS). IGI Publ. **8**(4), 23–42 (2017)
6. P. Dialani, Improved decisions through decision intelligence, 2020
7. K. Dear, Artificial intelligence and decision-making. RUSI J. **164**(5–6), 18–25 (2019)
8. E. Gürel, M. Tat, SWOT analysis: A theoretical review. J. Int. Soc. Res. **10**(51), 994–1006 (2017)

9. A. Khatri, T. Choudhury, T.P. Singh, M. Shamoon, Video analytics based identification and tracking in Smart spaces, in 2019 International Conference on Contemporary Computing and Informatics (IC3I), 2019, pp. 261–267
10. A. Gupta, A. Singhal, A.S. Sabitha, T. Choudhury, Water utility management using data analytics and geospatial analytics for state of Delhi, in Proceedings of the 2nd International Conference on Green Computing and Internet of Things, ICGCIoT, 2018
11. T. Choudhury, A.S. Chhabra, P. Kumar, S. Sharma, A recent trends on big data analytics, in 2016 International Conference System Modeling & Advancement in Research Trends (SMART), 2016, pp. 225–231
12. F. Kitsios, M. Kamariotou, M.A. Talias, Corporate sustainability strategies and decision support methods: A bibliometric analysis. Sustainability **12**(2), 521–542 (2020)
13. Analytics—What it is and why it matters. https://www.sas.com/en_us/insights/analytics/what-is-analytics.html. Accessed 9 Dec 2020
14. Why business analytics is so important for success. https://exeedcollege.com/blog/why-business-analytics-is-so-important-for-success. Accessed 9 Dec 2020
15. F. Diana, The evolving role of business analytics, 2011. Frank Diana's blog–our emerging future. https://frankdiana.net/2011/03/19/the-evolving-role-of-business-analytics. Accessed 9 Aug 2020
16. I.A. Ajah, H.F. Nweke, Big data and business analytics: Trends, platforms, success factors, and applications. Big Data Cogn. Comput. **3**(2), 32 (2019)
17. J. Frankenfield Descriptive Analytics, 2019. https://www.investopedia.com/terms/d/descriptive-analytics.asp#:~:text=Descriptive%20analytics%20is%20an%20important, intelligence%20a%20company%20will%20use. Accessed 10 Dec 2020
18. A.K. Sharma, S.P. Khandait, A novel software tool to generate customer needs tor effective design of online shopping websites. Int. J. Inform. Technol. Comp. Sci. **83**, 85–92 (2016)
19. Venkatesh. Decision-making techniques: traditional and modern techniques. https://www.yourarticlelibrary.com/management/decision-making-management/decision-making-techniques-traditional-and-modern-techniques/53215%23:~:text=It%20is%20method%20 %20of%20applying,involved%20in%20making%20various%20decision. Accessed 5 Aug 2020
20. SolveXia Blog, Improve business success with descriptive analytics, 2019. https://www.solvexia.com/blog/improve-business-success-with-descriptive-analytics. Accessed 10 Dec 2020
21. A. Bekker, 4 types of data analytics to improve decision-making, 2019. https://www.scnsoft.com/blog/4-types-of-data-analytics. Accessed 12 Dec 2020
22. S. Parthasarathy, What Is predictive analytics? 2020. https://www.logianalytics.com/predictive-analytics/what-is-predictive-analytics. Accessed 12 Dec 2020
23. Predictive analytics—What it is and why it matters. https://www.sas.com/en_in/insights/analytics/predictive-analytics.html. Accessed 12 Dec 2020
24. P. Juneja, What is predictive analytics and its importance to businesses. https://www.managementstudyguide.com/predictive-analytics.htm. Accessed 12 Dec 2020
25. J. Edwards What is predictive analytics? Transforming data into future insights, 2019. https://www.cio.com/article/3273114/what-is-predictive-analytics-transforming-data-into- future-insights.html. Accessed 12 Dec 2020
26. J. Pryce, A. Yelick, Y. Zhang, Using artificial intelligence, machine learning, and predictive analytics in decision-making, 2018, in White Paper Florida Institute for Child Welfare, pp. 1–9
27. K. Surendro, Predictive analytics for predicting customer behavior, in IEEE international conference of artificial intelligence and information technology (ICAIIT), 2019, pp. 230–233
28. S. Parthasarathy. What is predictive analytics? (2020), https://www.logianalytics.com/predictive-analytics/what-is-predictive-analytics/. Accessed 09 Dec 2020
29. A. Patrizio, Top predictive analytics examples: analytics for business success, 2019. https://www.datamation.com/big-data/predictive-analytics-examples.html. Accessed 9 Dec 2020

30. K. Jensen, The strategic impact of predictive analytics, 2015. https://www.ibmbigdatahub.com/blog/strategic-impact-predictive-analytics. Accessed 16 Dec 2020
31. P. Juneja, What is predictive analytics and its importance to businesses. https://www.managementstudyguide.com/predictive-analytics.htm. Accessed 17 Dec 2020
32. What is prescriptive analytics? https://www.sisense.com/glossary/prescriptive-analytics/. Accessed 12 Dec 2020
33. Troy Segal, Prescriptive analytics, 2019. http://investopedia.com/terms/p/prescriptive-analytics.asp#:~:text=Prescriptive%20analytics%20makes%20use%20of,to%20determine%20near-term%20outcomes. Accessed 17 Dec 2020
34. M. Walker, Prescriptive analytics strategy, 2015. http://www.datascienceassn.org/content/prescriptive-analytics-strategy#:~:text=To%20design%20and%20implement%20an,and%20a%20data%20science%20strategy. Accessed 17 Dec 2020
35. E. Roy, 4 types of data analytics to improve decision-making, 2019. https://www.knowledgehut.com/blog/big-data/4-types-of-data-analytics-to-improve-decision-making. Accessed 14 Dec 2020
36. Master of Business Administration. Predictive vs. prescriptive analytics: what's the difference? 2020. https://onlinemasters.ohio.edu/blog/predictive-vs-prescriptive-analytics-whats-the-difference/. Accessed 18 Dec 2020
37. K. Folse, Examples of prescriptive analytics. https://accent-technologies.com/2020/06/18/examples-of-prescriptive-analytics/. Accessed 17 Dec 2020

Chapter 2
A Complete Overview of Analytics Techniques: Descriptive, Predictive, and Prescriptive

Debashish Roy, Rajeev Srivastava, Mansi Jat, and Mustafa Said Karaca

2.1 Introduction

This is the era of digitalization. Anything and everything is moving toward becoming virtual and online. The data is growing exponentially every millisecond. Big data is taking the world by storm through its infinite number of benefits. All this has led to analytics becoming a major part of all the business processes in almost all the sectors around the globe. Let us understand what analytics is all about.

Analytics is the science or art of applying statistics, data mining, and technology to huge chunks of data for obtaining insights and actionable information. Analytics is a multidimensional discipline which involves data analysis and its visualization along with mathematics/statistics and machine learning. The ultimate objective or aim of analytics is effective decision-making by observing patterns of variations. While studying the patterns, the firms understand and predict uncertainties related to their business. Various techniques and models are used like statistical, economical, or mathematical models. The overall gist of analytics can be broken down into three Os of analytics (Fig. 2.1). Those are:

- *Obtain*: Obtaining or gathering or collecting the data is the first step in analytics. This is the most important and crucial part of the overall process. One should collect data based on the problem statement he/she needs solution for. As an example, we can talk about an entity collecting data from social media platforms to know how the users engage with their company.

D. Roy (✉) · R. Srivastava · M. Jat
School of Business, University of Petroleum and Energy Studies, Dehradun, India
e-mail: rsrivastava@ddn.upes.ac.in

M. S. Karaca
Vakif Katilim Bankasi A.S., Vakıf Katılım Bankası, Istanbul, Turkey

P. M. Jeyanthi et al. (eds.), *Decision Intelligence Analytics and the Implementation of Strategic Business Management*, EAI/Springer Innovations in Communication and Computing, https://doi.org/10.1007/978-3-030-82763-2_2

Fig. 2.1 Three Os of analytics

- *Organize*: After the data is collected, the next step is to organize or consolidate the data to make it more usable for analysis. It is the most logical step when we talk about analytics as different persons will be using that data for different purposes. So it has to be arranged in an orderly fashion so as to extract as much insights as possible.
- *Observe*: After the data is structured and organized, the next step is its analysis. Through analysis, we get to observe the relationships, patterns, and trends with the help of which we can predict the future outcomes.

Analytics in itself is a broader term that can be divided into three types:

- *Descriptive analytics*: what has happened in past?
- *Predictive analytics*: what is most likely to happen in future?
- *Prescriptive analytics*: what can be done for it?

2.2 Descriptive Analytics

Mankind has been observing the past since the times immemorial. Looking into the past gives a clear understanding of how things have led to the situation that is the present. Without observing past, mankind would have become comfortable with the present state of affairs. However, observing the past is one thing and measuring it is another. This act of measuring the historical data gave birth to the practice of descriptive analytics.

Descriptive analytics plays on the grounds of the understanding of past data to get a better perspective on the differences. Descriptive analytics elaborates on the utilization of data from a period of time to get conclusions.

This practice utilizes data from a time period, that is, a range to give a correct scenario of what has occurred and how it is in contrast with the other periods. This practice unveils the strong and weak areas of the organization. This further contributes in strategy development.

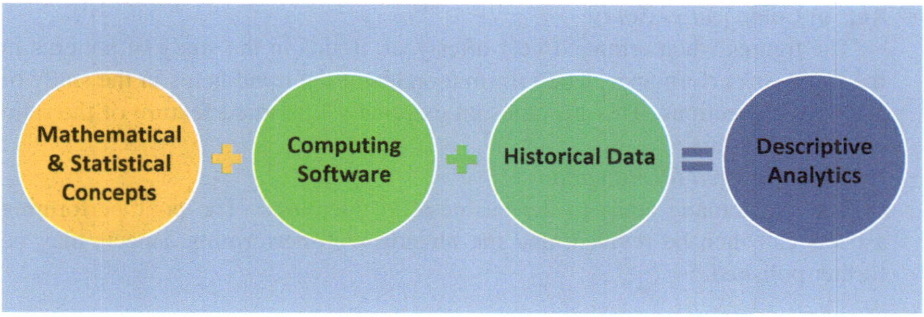

Fig. 2.2 Components of descriptive analytics

Descriptive analytics forms the basis of business intelligence, and it is profusely used by companies today. The practice of descriptive analytics can slightly differ from industry to industry. Though there are certain calculations without which the health of the business cannot be ascertained. These are financial calculations for which advanced software are not required unless the volume of the data is big.

This type of analytics provides for the management a visually easy-to-understand format to understand the data. Though the focus of industries is now shifting more toward the areas of analytics like the prescriptive or diagnostic analytics. For analytics like the descriptive, prescriptive, predictive, or the diagnostic, they use the descriptive analytics, and the addition of further data takes place to generate the likely outcomes or to find the reasons behind them (Fig. 2.2).

2.2.1 *The Ratio Analysis: An Elaborate Example of Descriptive Analytics in Business*

Calculation of accounting ratios is one of the best, widely used, and a very old practice of describing the financial health of a business. It describes the true and exact areas where a business lacks strength or gains strength. It describes exactly where the company is winning or failing.

Following are some examples of ratio analysis as a part of descriptive analytics:

- *Understanding in the Efficiency of Decisions*

  It helps understand if the business management is taking the correct decisions with respect to financing, operating, and investing.
- *Relationship Establishment and Simplification of Numbers*

  These ratios help in understanding the relationships between the financial elements of the business in an easier format. It gives the better understanding of the managerial effectiveness, the credit in the business, and the capacity to earn.

- *Aids in Comparative Study*

  The figures when arranged year-after-year, it aids in the study of patterns in the business performance. The information about the trend helps in the study of financial projections. This becomes an extremely important feature of the ratio analysis.
- *Problematic Area Identification*

  The problematic areas in the business are identified. The poor performing aspects can then be revived, and the already well-performing aspects may be further polished.

Types of Ratio Analysis

- Liquidity ratios
  - Current ratio
  - Quick ratio
- Solvency ratios
  - Debt-equity ratio
  - Debt to capital employed ratio
  - Proprietary ratio
  - Total assets to debt ratio
  - Interest coverage ratio
- Turnover ratio
  - Inventory turnover
  - Trade receivable turnover
  - Trade payable turnover
  - Investment (net assets) turnover
  - Fixed assets turnover
  - Working capital turnover
- Profitability ratios
  - Gross profit ratio
  - Operating ratio
  - Operating profit ratio
  - Net profit ratio
  - Return on investment or return on capital employed
  - Return on net worth
  - Earnings per share
  - Book value per share
  - Dividend pay-out ratio
  - Price earnings ratio

The analysis of certain ratios have been given below as an example:

$$\text{Current ratio} = \text{current assets}/\text{current liabilities}$$

	APOLLO 2018	APOLLO 2O19	APOLLO 2020	MAX 2020
Current ratio	1.51	1.23	1.19	0.88

Analysis

- APOLLO has current ratios throughout all the years above 1, which is a good sign.
- However, the falling current ratio of APOLLO is alarming though it is still above 1 which means that APOLLO has enough current assets to cover their current liabilities. The ideal current ratio for any organization should be from 1.2 to 1.5.
- MAX has a current ratio below 1; this means that MAX does not have enough current assets to cover its current liabilities. Therefore, though APOLLO is falling in its current ratio, it is still in a better position than MAX.

$$\text{Quick ratio} = \text{quick current assets}/\text{current liabilities}$$

	APOLLO 2018	APOLLO 2019	APOLLO 2020	MAX 2020
Quick ratio	1.10	0.88	0.83	0.84

Analysis

$$\text{Quick ratio} = (\text{current assets–inventory})\ /\text{current liabilities}$$

- It is clearly visible with APOLLO's performance on the quick ratio has fallen below 1 in 2019 and 2020.
- This is a bad indication for APOLLO's financial health. Since quick assets is not taking into account the inventory, it means that APOLLO's standing was still better without taking into account inventory in 2018.
- APOLLO clearly does not have enough liquid assets to cover the current liabilities.
- In comparison to MAX, one can say that it is clearly at an equally bad financial position with APOLLO when it comes to covering of current liabilities with the help of quick assets or liquid assets.

$$\text{Return on assets} = \text{net income}/\text{average total assets}$$

	APOLLO 2018	APOLLO 2019	APOLLO 2020	MAX 2020
Return on assets	2.98%	3.59%	4.63%	1.11%

Analysis

- This ratio is an indicator of how well an organization is using its assets to generate profits.
- The ROA of 5% is a good indicator and APOLLO is reaching it over the years. This means that it is improving upon its ability to generate more profits through its assets. APOLLO has reached 4.63% in 2020.
- MAX as compared to APOLLO does not as efficiently get profits as per the assets it has. APOLLO is at a better position than MAX in 2020.

$$\text{Return on equity} = \text{net income}/\text{average stockholder equity}$$

	APOLLO 2018	APOLLO 2019	APOLLO 2020	Max 2020
Return on equity	6.31%	7.79%	11.79%	2.23%

Analysis

- ROE of APPOLLO, over the years, is moving toward the percentage of 15%, and this is a good sign. A percentage of 15-20% is considered good for an organization's ROE. It means that the organization is efficiently utilizing the shareholder's money that is invested.
- MAX however lags way behind than APOLLO in the ROE in 2020 since APOLLO is at 11.79 and MAX is at 2.23%.

2.2.2 *The Gist of Descriptive Analytics*

As seen in the financial ratio example above, it can be said that descriptive analytics is not only in the advanced statistical tools or models. It may be present even in the simplest of calculations. All it needs is the right perspective to analyze it or rather describe it. This type of analytics maybe least in value as compared to others; however, it forms the basics of analytics and is important to understand the ropes of it. One can say that if one has to become a successful business analytics professionals, then descriptive analytics is the stepping stone.

2.3 Predictive Analytics

Predictive analytics is another category of data analytics whose aim is to make predictions about future outcomes taking historical data as a base. It is an area or domain of advanced analytics that many experts and organizations use to make predictions about the future events. To analyze the current data and make predictions, it uses some of the techniques mentioned below (Fig. 2.3):

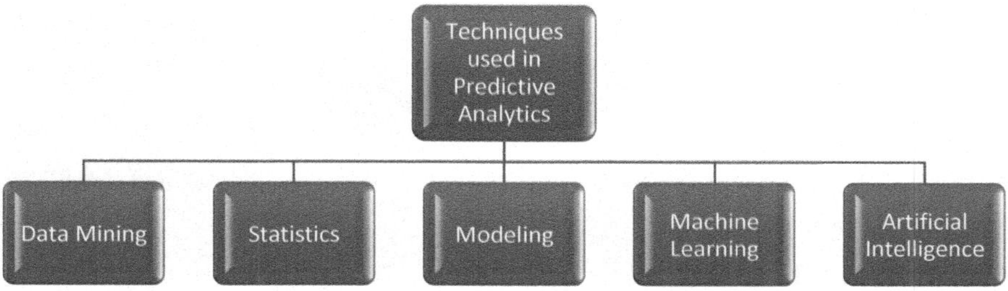

Fig. 2.3 Techniques used in predictive analytics

- Data mining

2.4 Statistics

- Modeling
- Machine learning
- Artificial intelligence

With the help of predictive analytics, future insights can be generated with a notable degree of precision. Effective and comprehensive predictive analytics techniques and tools can use the historical as well as current data to forecast trends in milliseconds, days, or years into the future. It also helps in bringing together the three processes, i.e., the management, information technology, and modeling to predict future events. The question here arises as to what to do with those predictions. Well, the patterns generated by the historical and current data help in identifying two crucial aspects of the businesses:

- Risks
- Opportunities

The predictive analytics models try to build up a relationship among the factors or variables that help in assessing the risks in a particular set of conditions by assigning a particular score or weightage to those factors. Successful and comprehensive implementation of this branch of analytics can help the firms use big data to extract as much benefit as they can. The organizations, with the help of predictive analytics, become pro in every manner. In other words, it can be said that they become proactive and progressive in every business process they implement in their business by uncovering patterns and relationships in both the structured and unstructured data. The structured data here can be referred to as data that is ready for use and implementation of various tools and techniques, whereas unstructured data is the one that needs more cleaning and processing.

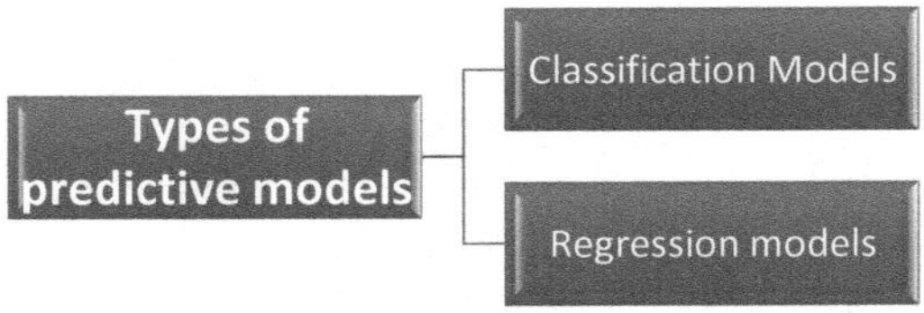

Fig. 2.4 Types of predictive models

2.4.1 How Does It Work?

The models developed using predictive analytics use the results to train a model which can be used to predict values for a new data set. The models provide results in the form of predictions representing a probability of the variable targeted (e.g., sales) based on significance estimation from a set of input or the independent variables (advertising costs, investment, etc.).

Let us understand this more clearly. There are basically two types of predictive models (Fig. 2.4).

2.4.1.1 Classification Models

These models help predict class label. For example, predicting whether the match will happen or not, whether someone is likely to leave the organization or not, whether his credit score is good or bad, etc. The results generated by this model are usually in the form of 0 or 1, with 1 as the event we target. Classification models include decision tree, logistic regression, random forest, gradient-boosted tree, etc.

2.4.1.2 Regression Models

These models help predict a number (quantifiable variable). It is a predictive modeling technique that helps in investigating the relationship between a dependent variable and independent variable(s). Predicting how much revenue a company would generate next year, how much CO_2 will be there in the atmosphere next year are some of the examples that can be considered here.

2.4.2 Predictive Analytics Process

- *Fix*: It is very important to be specific about what you wish to achieve with the implementation of predictive analytics. The projects and the objectives need to be fixed and predefined so as to make the process more effective and easy. The outcomes, deliverables, as well as the input which would be used needs to fixed in the beginning itself.

- *Fetch*: Fetching here refers to all about the collection of data from multiple sources by both or either of the data collection methods, i.e., primary or secondary. Predictive analytics is all about the use of large volumes of data to generate insights about the trends and stay one step ahead in the game. Therefore, the data collection phase is the crucial one for the successful implementation of the process. The data here would revolve around the information from multiple sources but with a unitary approach to it.
- *Analyze*: After fetching the data, it is time to anatomize it. This would lead to revelation of trends to reduce the risks and optimize the processes. This step up to a larger extent involves cleaning and structuring of data rather than modeling. Statistics here is another important aspect to be paid attention to for the testing and validating of assumptions. Those in charge of the projects would define a specific hypothesis and test it using statistical techniques to make decisions based on numbers.
- *Build*: The next step in the process is to build the model. When talking about modeling, the tool to be used plays crucial role. There are innumerable libraries in the open-source programming languages which include Python and R. In this stage, the accurate models about the future are developed by choosing the best option from the various options available.
- *Deploy*: After going through the process of statistical analysis and model building, it is necessary to interpret and integrate the outcomes into daily routine activities. It should become a key factor in making the routine choices and developing the processes in the firms. By this, we can conclude, having the numbers that show what is best for the business is not enough. It is important to translate those outcomes into actionable and measurable steps.
- *Monitor*: Nothing in this world can remain static be it the case with data, tools, techniques, or the conditions. A model's validity would remain for a particular time period that is till the time the external environment and conditions do not change notably. Therefore, revisiting and testing the models with new data and ensuring its significance would count as a good practice to stay ahead in the game.

A better clarity on the process can be fetched with the pictorial representation in Fig. 2.5.

2.4.3 Predictive Analytics Tools

The array of business activities is endless. The tools used for predictive analytics give users deep and real-time insights into that endless array. Based on the analysis of data collected over a period of time, the tools can predict various types of behavior, trends, and patterns. For example, how to allocate the resources, when

Fig. 2.5 Predictive analytics process

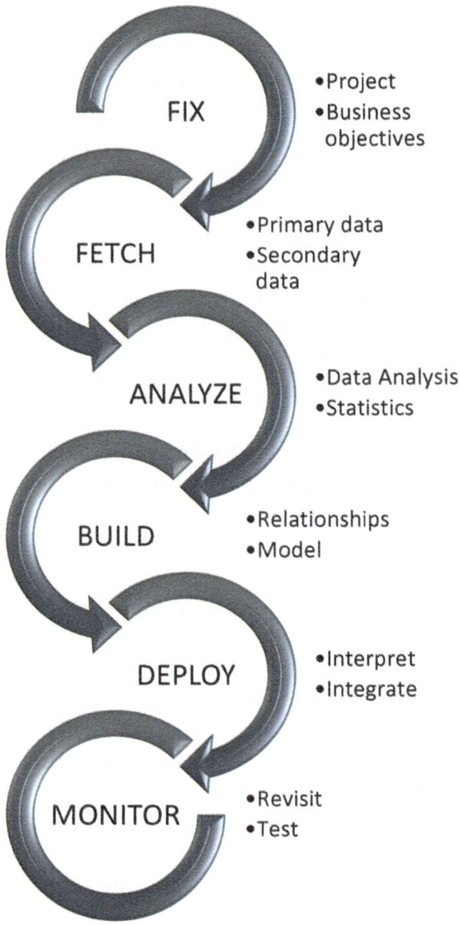

is the best time to launch a product or marketing campaign, etc. Some of the major predictive analytics software and the service providers are listed below:

- IBM
- Microsoft
- SAP
- SAS Institute
- Tableau Software

2.4.4 Uses of Predictive Analytics in Different Industries

Almost every industry today is using predictive analytics for reducing the risks, optimizing the operational activities, and increasing revenue. Some of the examples are listed below:

- *Banking and financial sector*: We all know financial sector involves huge amount of money at stake along with big chunk of data. This sector has adopted predictive analytics way back for reduction in frauds, measurement of risks related to credit, maximization of cross-sell or up-sell opportunities and most importantly to retain its valuable customers. Predictive analytics lets the firm develop credit risk models, forecast the trends in financial market, predict the impact of new policies on their businesses as well as the markets.
- *Retail sector*: The retail sector all around the world is using predictive analytics for planning merchandise and optimizing the price. The biggest and relevant example can be of Amazon which automatically provides the users with combinations of products the customers would be most likely to buy. The retailers also use predictive analytics for analyzing how effective their promotional events are. It helps them to determine which offers are most comprehensive and appropriate for the customers.
- *Oil and gas sector*: Whether it is reducing the safety risks, predicting the equipment failures or future needs of resources, or improving the overall performance, the energy sector has adopted predictive analytics with utmost robustness. The need of maintenance of huge power-generating turbines is predicted using machine sensors data analysis.
- *Healthcare sector*: The use of predictive analytics in healthcare sector is helping in detecting early signs of deterioration of patients in the ICU and general wards. It is also benefiting the healthcare sector by identifying the patients at risk in their homes that helps in preventing the downtime of medical equipment and services. Apart from all this, hospitals even use predictive analytics in tackling and reducing the challenges related to operations and administration.
- *Manufacturing sector*: Lenovo, a Chinese multinational technology company, has used predictive analytics for better clarity and understanding of the warranty claims. This step taken by them led to approximately 15% reduction in their warranty costs. This is just an example. The manufacturers around the globe are using predictive analytics for the identification of the factors leading to the poor quality or failures in production activities, etc. It is also helping them in optimization of the resources and distribution.

2.4.5 Why Now?

Predictive analytics has been around for decades but its usefulness has come of age as a core practice that is crucial to sustain in today's market. It empowers the businesses by applying organizational learning to grow exponentially by deploying a unique form of data-driven risk management system. With interactive and easy-to-use software becoming more prevalent, predictive analytics is no longer just the domain of mathematicians and statisticians. Business analysts and line-of-business experts are also using these technologies to gain insights from data-driven techniques. Why has it become so relevant today?

- The volume of data is growing exponentially day by day. People or firms are building high interest in using data to generate valuable insights.
- Today's economic conditions are tougher which has generated a need for competitive differentiation and advantage.
- The computers available today are faster, cheaper, and in various sizes and models.
- The software available are easier to use.

2.5 Prescriptive Analytics

2.5.1 Introduction

Business analytics uses a large amount of data collected from different sources, apply quantitative and statistical analysis to develop predictive models to help different stakeholders in decision-making [6, 19].

It can be categorized into three main stages, namely descriptive, predictive, and prescriptive analytics. These categories are based on the type of difficulty, value, and intelligence level [1, 12, 18]. Descriptive analytics is mainly used to answer the questions like "what happened," "why it happened," or "what is happening." Predictive analytics is used to answer questions like "what will happen" and "why will it happen." Similarly, prescriptive analytics is used to answer questions like "what should we do" and "why should we do it."

It has been found that still, the major focus of business analytics is on descriptive and predictive analytics using methodologies like machine learning, artificial intelligence, etc. [7, 10, 13]. In comparison with descriptive and predictive analytics, prescriptive analytics is less mature [9]. In this chapter, the background of prescriptive analytics is discussed along with the categories of various methodologies used in prescriptive analytics.

2.5.2 Background of Prescriptive Analytics

Prescriptive analytics can be used to get more value and intelligence to the business [18]. Using prescriptive analytics, one can get the best decision options among the options received through predictive analytics. There are two levels of human interaction in prescriptive analytics decision support and decision automation. The decision support example is giving recommendations, and the decision automation example is the implementation of prescribed action [9]. How effective the predictive models are is dependent on the type of structured and unstructured data, the domain of the study, and finally the impact of analysis [3, 18].

2.5.3 Methods for Prescriptive Analytics

Various types of methods are used for prescriptive analytics. These methods can be divided into six main categories which are shown in Fig. 2.1. The names of these seven categories are machine learning (ML), mathematical programming, probabilistic models, simulation, evolutionary computation, and logic-based models. The details of these methodologies are given in this section (Fig. 2.6).

2.5.3.1 Probabilistic Models

This model is used to measure the uncertainty by using the process of integration of data with the first principle knowledge [14, 15].

In this category, the models represent the uncertain cause and effects relationship. In prescriptive analytics, these types of models are used to find the likelihood of the events.

2.5.3.2 Machine Learning

Machine learning (ML) learns from the historical data refer to an algorithm without using clear instructions [16]. ML algorithms used training data to develop a model and test data to validate a model. It is also known as the subset of artificial intelligence. Data mining is used to find a pattern in the huge data set to get insight into the data [4]. Data mining and ML terms are closely related to each other. By using predictive analytics, one can predict the future outcome by looking at

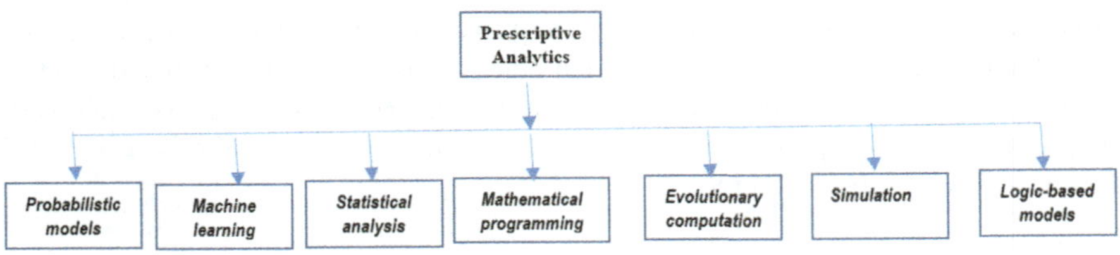

Fig. 2.6 Prescriptive analytics methods

the pattern found in the previous year or historical data. While, using prescriptive analytics can be helpful to find the best solution for the given problem.

2.5.3.3 Statistical Analysis

The branch of mathematics that deals with the collection of data, its organization, analysis, interpretation, and finally, presentation is known as statistical analysis [8, 17]. This analysis deals with the different aspects of the data which include a method of data collection, experiment design, and finally solve problems using statistical populations [17]. In predictive and prescriptive analytics, this analysis is useful to get insight from the data and use it to predict behavior and trends.

2.5.3.4 Mathematical Programming

Mathematical programming can be defined as the process of optimally allocated the limited resources among the competitive activities under certain constraints that are considered under study. It can also be defined as the branch of management science, mathematics, and operation research used to give better decision-making for complex problems [5].

2.5.3.5 Evolutionary Computation

It is a subset of AI and soft computing and is considered as a family of the algorithm used for global optimization [2]. This method is used for prescriptive analytics to solve complex problems in the data-rich environment where exact solutions are not possible.

2.5.3.6 Simulation

This process is used for modeling real-life situations on the computer platform to understand the performance of the system [11]. By changing the configuration of the model using expert knowledge one can easily predict the behavior of the system which facilitates decision-making. In prescriptive analytics, simulation can be used for effective decision-making by a human being. This method is very useful to test the new ideas used in business decisions to reduce the risk at various levels and very helpful to modify the processes. In prescriptive analytics, this method is very useful for the safety of infrastructure.

2.5.3.7 Logic-Based Models

This method is used to solve problems where there is a sequence of causes and effects to reach some solution. It is based on the rule-based system, which requires expert and domain knowledge for better decision-making used in the prescriptive analytics application.

2.6 Conclusion

Prescriptive analytics is seen as a critical advancement in the field of analytics. It helps an analyst to make effective decision-making by selecting the best option. It is less popular as compared to descriptive and predictive analytics and is applicable in some specific domains. Different types of methods are used in prescriptive analytics which are broadly divided into seven categories. These categories are machine learning (ML), mathematical programming, probabilistic models, simulation, evolutionary computation, and logic-based models.

In the future, prescriptive analytics becomes more common and applied by a vast variety of managers. After getting business value from the huge volume and variety of data, it requires quick action especially in the case of real-time events. Taking quick and intelligent action using prescriptive analytics requires more than predictive analytics.

2.7 Conclusion

Businesses all around the world are utilizing the data to discover valuable insights to make their decision-making process effective and efficient. The usefulness and the functionality of all the three distinct types of analytics—descriptive, predictive, and prescriptive—is vast. They can be used individually or as a combination depending on the case-to-case basis to make business a flourishing one. As more and more firms are realizing the importance of big data as a competitive advantage, they must ensure to choose the right kind of analytics technique or combination of any two or all the three to lead their paths toward required solutions.

References

1. B. Akerkar, Advanced data analytics for business, in *Big data computing*, (CRC Press, Boca Raton, 2013), pp. 373–397
2. T. Bäck, D.B. Fogel, Z. Michalewicz, *Handbook of Evolutionary Computation* (CRC Press, Boca Raton, 1997)

3. A. Basu, Five pillars of prescriptive analytics success. Anal, Magaz **8**, 8–12 (2013)
4. S. Chakrabarti, M. Ester, U. Fayyad, J. Gehrke, J. Han, S. Morishita, W. Wang, Data mining curriculum: A proposal (version 1.0), 2006. Intensive Working Group of ACM SIGKDD Curriculum Committee, p. 140
5. E.K.P. Chong, S.H. Zak, *An Introduction to Optimization* (Wiley, New York, 2008)
6. T.H. Davenport, J.G. Harris, *Competing on Analytics: The New Science of Winning* (Harvard Business Press, Boston, 2007)
7. D. Den Hertog, K. Postek, Bridging the gap between predictive and prescriptive analytics-new optimization methodology needed, 2016. Technical report, Tilburg University. http://www.optimization-online.org/DB_HTML/2016/12/5779.html
8. Y. Dodge, *The Oxford Dictionary of Statistical Terms* (Oxford University Press, Oxford, 2006)
9. Gartner, Planning guide for data and analytics, 2017. https:// www.gartner.com/binaries/content/assets/events/keywords/catalyst/catus8/2017_planning_guide_for_data_analytics.pdf, Accessed 3 April 2018
10. R.A.A. Habeeb, F. Nasaruddin, A. Gani, I.A.T. Hashem, E. Ahmed, M. Imran, Real-time big data processing for anomaly detection: A survey. Int. J. Inf. Manag. **45**, 289–307 (2018)
11. B. Jerry, *Discrete Event System Simulation* (Pearson Education, New Delhi, 2005)
12. J. Krumeich, D. Werth, P. Loos, Prescriptive control of business processes. Bus. Inf. Syst. Eng. **58**(4), 261–280 (2016)
13. D.T. Larose, C.D. Larose, *Data Mining and Predictive Analytics* (Wiley, New York, 2015)
14. E.C. Martinez, M.D. Cristaldi, R.J. Grau, Design of dynamic experiments in modeling for optimization of batch processes. Ind. Eng. Chem. Res. **48**(7), 3453–3465 (2009)
15. E.C. Martínez, M.D. Cristaldi, R.J. Grau, Dynamic optimization of bioreactors using probabilistic tendency models and Bayesian active learning. Comput. Chem. Eng. **49**, 37–49 (2013)
16. N.M. Nasrabadi, Pattern recognition and machine learning. J, Electr, Imag. **16**(4), 049901 (2007)
17. J.W. Romijn, Philosophy of statistics, in *Stanford Encyclopedia of Philosophy*, (Stanford University, Stanford, 2014)
18. L. Šikšnys, T.B. Pedersen, Prescriptive analytics, in *Encyclopedia of Database Systems*, ed. by L. Liu, M. Özsu, (Springer, New York, NY, 2016)
19. R. Soltanpoor, T. Sellis, Prescriptive analytics for big data, in *Databases theory and applications*, ed. by M. A. Cheema, W. Zhang, L. Chang, (Springer, Sydney, NSW, 2016), pp. 245–325

Chapter 3
Artificial Intelligence and Analytics for Better Decision-Making and Strategy Management

Neelaksh Sheel, Lal Pratap Verma, Somesh Kumar, and T P Singh

3.1 Introduction

Presently, AI is the only reigning technology revolution of human life, with the potential capacity to disturb more or less all the features of human nature. "Co-founder of Coursera and previous leader of Baidu AI Group / Google Brain, Andrew Ng" evaluate the conversion effect of AI to that of electrical energy after 100 years. As a lot of industries violently investing in AI solutions, international investors anticipate to attain a "Compound Annual Growth Rate (CAGR) of 50.1% to reach $57.6 billion in 2021" [1]. Artificial intelligence is not a novel concept nowadays, a number of hypothetical and technological keystones have been developed since last 70 years. A number of computer scientists like Marvin Minsky, Alan Turing, and John McCarthy had previously worked upon them. Even AI has been already implemented in different government and private industries. Due to virtual unrestricted computation and declining cost of memory space, exponential growth of AI as organization finds out to unlock the values in huge volumes of data. AI is an assemblage of techniques that allow machines to take action with elevated heights of intelligence and replicate the human potential of understanding and perception. By recording and processing images, noise, and voice, computer vision can dynamically discern the environment. AI will analyze and understand

N. Sheel (✉) · S. Kumar
Moradabad Institute of Technology, Moradabad, India

L. P. Verma
Manipal University, Jaipur, India

T. P. Singh
Department of Informatics, University of Petroleum and Energy Studies, Dehradun, Uttarakhand, India
e-mail: thipendra@gmail.com

P. M. Jeyanthi et al. (eds.), *Decision Intelligence Analytics and the Implementation of Strategic Business Management*, EAI/Springer Innovations in Communication and Computing, https://doi.org/10.1007/978-3-030-82763-2_3

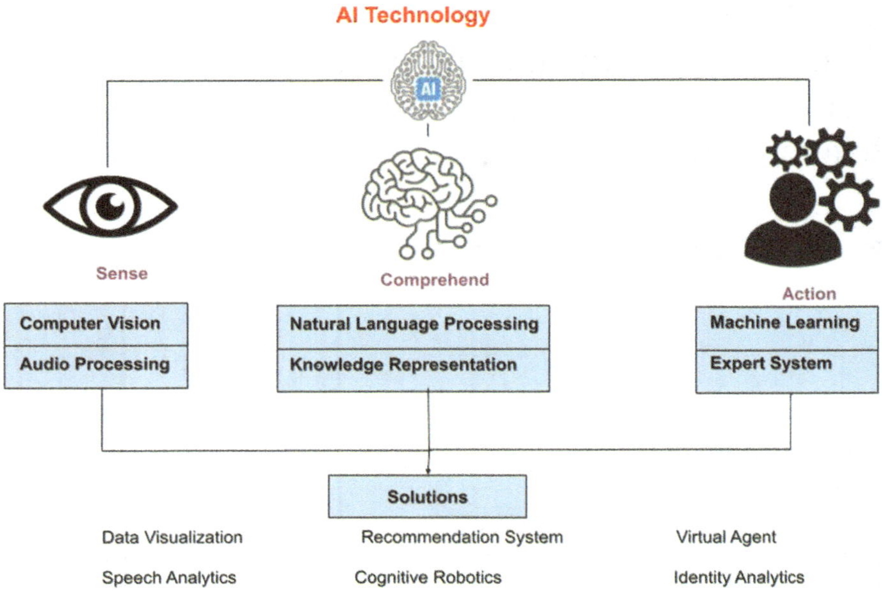

Fig. 3.1 Artificial intelligence

knowledge thanks to "Natural Language Processing (NLP)" and inference engines. Expert systems and inference engines, for example, consider actions from the real world, and AI computer implementations may take action all the way through experience. The ability to use previous experience and adapt it on a regular basis improves human capabilities. As AI applications become more sophisticated, they are constantly looking for large applications to append these capabilities around the world. Irrespective of the style of AI being used, every machine or applications initiate with large set of training data. Previously, arithmetic regressions, i.e., rules-based data analytics [2–4] applications or premature "expert systems", were used to extract these types of behavior. Deep learning and neural networks have now given us a simple but efficient AI program that is capable of doing something unexpected (Fig. 3.1).

Artificial intelligence has familiarized a great recitation of optimism followed by dissatisfaction, designated as "AI winters." Every preceding rift has moderately satisfied to the exposure it created, but no one has handled to initiate this technology to reach up to the conventional.

Today's revolution gives us an extraordinary phase of technological innovations across the world, due to which different fields give us the certainty that the "AI Spring" has not only appeared on the panel but will stay here for long time. Key expansion for AI optimism is:

1. *Unlimited Power of Computing*: As per the Gartner, Desktop as a Service (Daas), the globally cloud service marketplace is estimated to grow 6.3% by 2020 of

$257.9 billion, reached $242.7 billion till 2019. The admittance is increased as the computational power of computer is increased.

2. *Declination of storage data cost*: We are in a period where the cost of a hard disc drive per gigabyte has been falling at an exponential rate, to the point that we are approaching a near-zero marginal cost for data storage (from 500,000 USD per gigabyte in 1980 to 2 cents per gigabyte in 2017).
3. *Digitized Data*: According to the "International Data Corporation (IDC)," the international data bubble would rise to 163 ZB, or a trillion gigabytes, by 2025 [5] or ten times the data generated in 2016. According to Barry Smith, a professor in the Department of Computer Science at "University College Dublin," data is to AI what food is to humans. "As a result, in this digital age, the exponential increase in data feeds AI, allowing it to create more "intelligent machines".

In addition to increasing memory, we need faster processing units, such as "Graphics Processing Units (GPU's)" instead of "Central Processing Units (CPU's)." In May 2017, Google revealed that its "Tensor Processing Unit (TPU)" delivers more output per watt than today's CPUs and GPUs [6]. We need a specific and efficient combination of equipment with high data handling capabilities as the cost of storage decreases and the data volume grows exponentially. All of this adds up to a hugely influential foundation for AI in its substantial accumulation for traditional adoption.

AI gets categorized in diverse traditions and may be helpful to recognize the diverse classification, their rationale as well as suggestions.

1. *Weak and Strong AI*: Weak artificial intelligence describes simulated thought. Weak AI-based machines can attempt to mimic human actions and appear intelligent in doing so. It is completely unaware of what it is doing. For instance, a chatbot, Alexa, or Siri could materialize a popular conversation, but it has no understanding of who it is or why. When a powerful AI system employs a complex algorithm that aids the machines' actual thought, it is known as deep AI. Strong AI acts intelligently, thinks like a person, and has a subjective mind. For example, if the machine hears "Good Morning," it decides to turn on the coffee maker.
2. *Narrow and General AI*: Narrow AI can only operate on a single task at a time. Narrow AI communicates with humans on a single specified level. For instance, IBM's Deep Blue, a chess-playing machine, has the ability to beat world champion Gary Kasparov in 1997. This machine is only capable of playing chess. This machine is unaware of any other game strategies.
3. *Super Intelligence*: The phrase "super intelligence" is extensively used to explain power and general artificial intelligence that surpasses human intelligence. Although in Narrow Artificial Intelligence, a wide stride out has been achieved. Algorithms used to process documents, drive a vehicle in automatic mode, or play chess in real time with an opponent are not considered the first phase in artificial intelligence development.

Although advances in narrow artificial intelligence have been made, such as self-driving cars, document processing, and game play, no one has claimed to be the first to invent or create general AI. Professional opinion is that General AI is still a long way off from being a reality.

"AI technical advancements and technologies are rapidly evolving, with major implications for economies and humanity. According to a report conducted by EY and NASCCOM, by 2022, approximately 46% of workers will be working in entirely new positions that do not exist today, or in jobs that have dramatically changed skillets." [7]. If a few countries band together and plan to develop AI policies and lay the groundwork for the development of an AI ecosystem, they will be able to maintain the current momentum in the rapidly changing socioeconomic climate.

3.2 India and AI

National AI strategy needs to frame the policies and design a framework for AI, which should be adopted to fulfill India's sole requirements and ambition; on the other hand, it must be capable of accomplishing the country's intellectual potential of leveraging *Artificial Intelligence* expansion. This aforesaid framework could be perceived an accumulation of individual but interconnected components given below are:

1. *AI Opportunity*: For financial gain through AI
2. *AI for Goods*: Required for communal and comprehensive growth
3. *AI Global Market*: Provides the solution for the emerging technologies and raising economy throughout the globe

- *Opportunity: for financial gain through AI*

  AI holds promise as an innovative factor of invention, enhancing conventional factors of production such as labor, investment, and innovation, as well as technological changes reflected in major productivity factors. AI has a number of potential applications for overcoming financial and labor constraints, as well as unlocking new sources of value and development. In terms of finance, artificial intelligence has the ability to boost overall growth by allowing better decision-making:

  (a) *Innovation transmission*: AI innovations in energy sector like power sector, renewable energy sector, power trading; energy management will have optimistic consequences in many other fields. We know that industries are mutually dependent on value chain. Economic growth predicts from the new commodities, services, and modernism can be enabled through AI.
  (b) *Intellectual automation*: AI has the capability to automate sophisticated substantial tasks throughout the world that require adaptability and liveliness across industries.

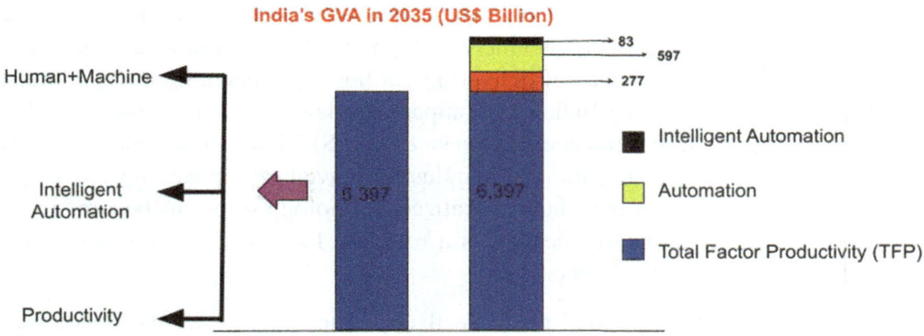

Fig. 3.2 Release innovation through AI

(c) *Labor and capital augmentation*: AI gives the opportunities to the researchers, scientists, and engineers to focus on various challenging career opportunities that enhances the human values and improves the economic efficiency of labor.

"Accenture, in its current AI research reports, present a framework for assessing the monetary impact of AI for select G20 nations and estimates AI to enhance India's annual growth rate by 1.3% age spots by 2035" [8] (Fig. 3.2).

- *AI for goods: required for communal and comprehensive growth*

  Adoption of disruptive technology such as artificial intelligence and machine learning needs to be seen as transformative impression on the society. It might be for the greater goods which improve the quality of life and admittance to a vast number of sections of the nation. With this wisdom, the latest advancements in AI seems to be conventional for the exceptional prospects and confronts that excellent healthcare facilities, comprehensive monetary growth in the large segment of inhabitants, providing real/on-time counseling to farmers. Examples are as follows:
- *AI global market*

  India, in addition to offering unrivalled opportunities, is an ideal recreational area for global business enterprises to create unique, scalable, and durable solutions for a variety of goods. In other emerging and promising economies, the product solution could be simple to implement. Based on technology, the solution can solve more than 40% of India's problems and can also help solve more than 40% of the world's problems. Other industries, such as crop farming, teaching, training, and transportation, are being used to bring the planet to the next level through AI. Public problems in specific industries in developed countries can be resolved by implementing AI solutions that are adaptable to different markets. As a result, AI technologies developed for the agricultural sector can be used in other developing countries based on their environment. In almost all developed countries, education will continue to be a major issue. Indians who speak a variety of languages will be able to receive high-quality education using AI-based technologies. They can also be very useful in countries that are still in the early stages of economic development. India's ability to lead AI is bolstered by its track record of delivering cutting-edge technological solutions. Making stuff in India (by Indian IT companies) may be a perfect model for the future of "Artificial Intelligence as a Service (AIaaS)." For technological product development and industry, India's IT industries have been opened up to the rest of the world. As AI matures and generalized technology solutions become more widely available, India will benefit as a result of large-scale development and implementation around the world.

Furthermore, proficiency in IT pooled with opportunities, such as interoperability among several languages, furnishes the desirable impact for adaptable algorithmic solutions that have global inference, such as Natural Language Processing.

3.3 Potential of AI

All the business leaders and investors available around the globe universally agree that AI has the great probability to transform their business by reducing cost, capable to manage risks, accelerate economic growth, and refueling innovation. AI provides large incremental values to a wide range of diverse industries around the world and is further expected to be the primary source of viable benefits for industries.

3.3.1 Healthcare

Healthcare is the biggest active and demanding sector, and the probability is to expand up to $280 billion by 2020, i.e., approximately 16% compound annual growth rate (CAGR) from the current scenario [9]. However, it has many difficulties of accessibility and affordability for large section of the residents.

"Insufficiency of competent medical proficient and services like qualified doctors, nurses and infrastructure, as confirmation in 0.76 doctors and 2.09 nurses per 1000 population (WHO recommendations of 1 doctor and 2.5 nurses per 1000 population respectively) and 1.3 hospital beds per 1000 population are available in India (WHO recommended 3.5 hospital beds per 1000 population" [10].

1. Cost efficiency remains an issue with personal expenses accounting for around 70% of healthcare operating cost, of which approximately 62% is out of conceal expenditure, almost certainly one of the uppermost in the world. As per government estimation, up to 63 million populations faced deficiency because of their healthcare expenditure [11].
2. Imprudent approach to vital healthcare mostly due to the short of responsiveness, lack of knowledge of how to avail services and incorrect lifestyle implies that the major number of patients move toward a hospital only when advance stage of

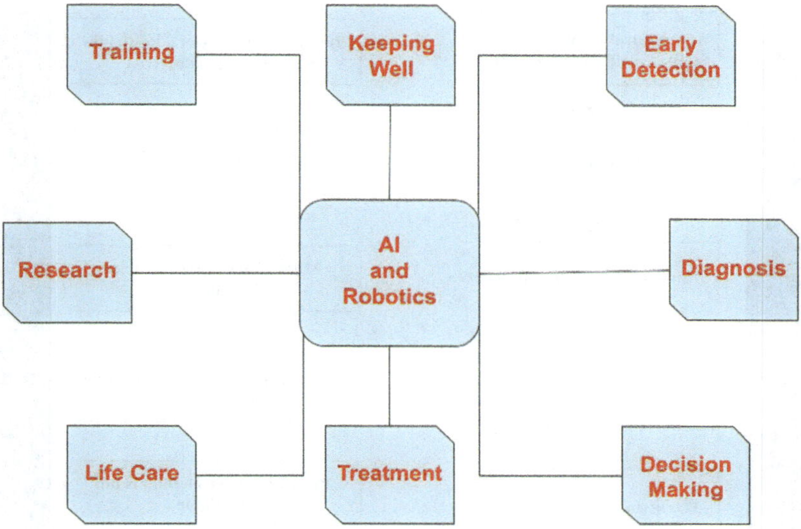

Fig. 3.3 Potential of AI in health care

diseases has reached; therefore, cost of medical care increases and the probability of recovery decreases.

Indian government has continuously trying to make a series of large-scale involvement to minimize the challenges of India's Healthcare, viz. "*Ayushman Bharat Mission*" promotes e-Health etc. (Fig. 3.3).

3.3.2 Agriculture

"Although India has come a long approach from being classified as purely an agricultural economy, agrarian and associated sector still accounts for 49% of India's labour force, 16% of the nation's gross domestic product (GDP)" [12], make sure the security of food for overall population.

Agriculture and its related industries are critical to India's growth. India's agriculture sector must produce at a rate of 4% or higher to achieve an annual growth rate of 8–10%. The Government of India has re-prioritized doubling farmer income as a National Agenda, putting a strong emphasis on agricultural supply chain growth and market development, as well as crop production enhancement. Despite making significant progress and attracting government attention, this sector remains reliant on impulsive variables such as supply chain and low productivity [13].

"India has not been able to fully break free from its exploitative reliance on resource-intensive agricultural practices. Reduced soil fertility, rapidly falling water tables, emerging pest resistance, increased reliance on inorganic fertilizers for increased production, and land degradation are just a few of India's indefensible agricultural practices" [14]. If the world's climate becomes more vulnerable, depen-

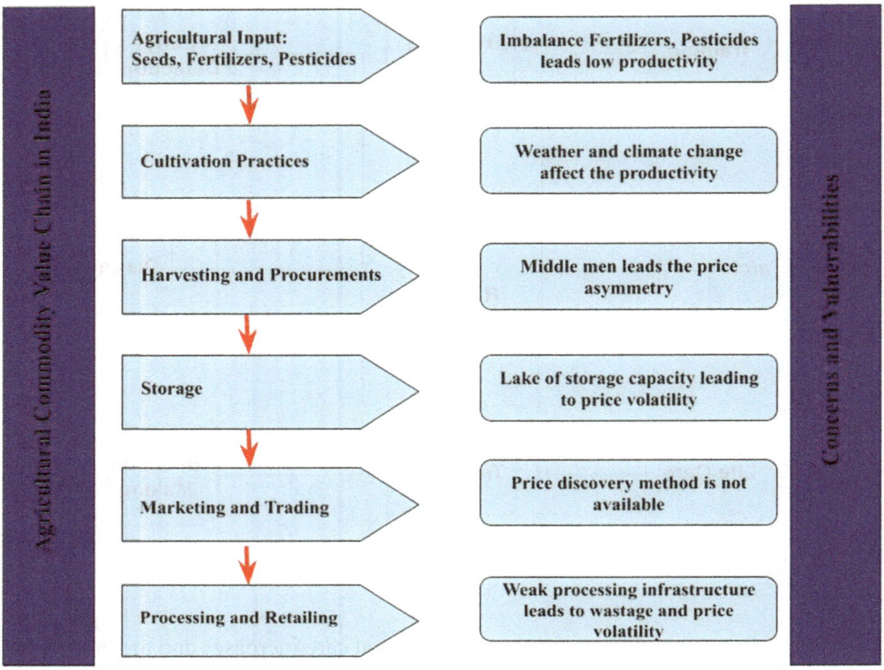

Fig. 3.4 Agri-commodity value chain

dence on non-sustainable and capital-intensive agriculture would only heighten the risks of food shortages and agricultural misery [14] (Fig. 3.4).

"AI will have major worldwide impact on productivity by agriculture at all levels of the value chain. An estimate by Markets and Markets Research valued AI in agriculture to be USD432 million in 2016 and expects it to increase at the rate of 22.5% CAGR to be valued at USD2.6 billion by 2025" [15].

According to "CB Insights" (a market intelligence platform), agricultural technical startups lift up over $800 million since last 5 years. "Deals for startups using automation and machine learning to resolve problems in agriculture initiate gaining momentum in 2014, in order with the increasing interest in AI transversely multiple industries akin to healthcare, finance, and commerce.". "From analyzing millions of satellite images to finding healthy strains of plant microbiome, these startups have elevated over 500 million USD to bring AI and robotics to agriculture" [16].

"In 2016, just about 50 Indian agricultural, technological startups ('AgTech') raised $ 313 million. This sector is considering extensive contribution through startups. *Intello Labs,* utilize image recognition software to supervise crops and forecast farm yields" [17]. "Aibono uses agricultural data science and AI to make available solutions to become stable crop yields." For monitoring the crops of farmers in real time, *Trithi Robotics* uses drone technology and gives accurate analysis of the soil. One more startup "*SatSure*" in India uses machine learning to review images of agricultural soil and forecast the monetary value of their future crops.

AI has the potential to increase the production and effectiveness on all given stages of the agricultural production series.

- Restoration and soil monitoring
- Crop monitoring and real-time advisories to farmers
- Increasing efficiency of farm mechanization
- Increase in price and its realization to the producer

3.3.3 *Education*

Everyone knows the importance of education; this sector is continuously augmented due to the large population of youngster in India. More than half of the country's population is expected to be under the age of 25 years. Acceptance in digital education means more data can be collected, so it is critical that these approaches are effectively implemented to enhance education and teaching.

Technology acceptance in today's educational system is growing, but not at the required pace. "Educational technology is expected to cost about $ 160 billion in schools around the world" [18]. However, low teacher retention rates and a weak educational effect have a negative impact on enrollment growth.

1. Low maintenance rates
2. Poor learning outcomes

Result of a compound interaction of the above two factors creates different test cases for the improvement in quality education:

1. *Empty post of teachers due to uneven locations*: Owing to an unequal distribution of location and accessibility issues, a substantial number of faculty vacancies are vacant or unfilled.
2. *Professional course/training does not meet the real needs and has low coverage*: The preparation that teachers have access to is usually a generic form of exercise. It has little to do with a teacher's individual demands or weaknesses; a teacher who does not understand arithmetic concepts requires corresponding training to explain arithmetic concepts. As a result, the majority of teacher training exercises are a waste of public funds, with little to no value to the teachers or their students. Other employees, such as lab attendants and principals, face similar issues with preparation. On an annual basis, fewer than 20% of training programs run, which is extremely low.
3. *Low recognition of technology*: According to a recent study, technology acceptance in schools is inadequate. Although the ICT infrastructure is adequate, due to a lack of teacher training at that stage, everything is a waste. Although 83% of teachers have computers, their use is restricted to audio/video and visual displays, as well as student practice. "For tracing student data participating in

forums, only 41% and 27% of technologies are used, respectively. This pattern is even more evident in the school segment that charges students a low fee" [19]. Some qualified teachers also enjoy using technology in the classroom. As per survey 88% trained teachers making use of computers for student online forum as compared to the 53% of untrained teachers [20].

Various tools of AI are being perfectly used over the globe and can be adopted in India also.

1. Online education tools for present education
2. Online problem-solving interactive system
3. Tools to notify preemptive action to give up school
4. Automated rationalization of teachers
5. Customized professional development courses

3.3.4 Smart Mobility in Transportation

"Transportation is the backbone of any nation. Society demands for the qualitative as well as quantitative mobility facility for peoples and goods" [21]. Transportation sector faces a variety of issues.

Some other issues are congestion and road accident, lack of public transportation infrastructure and causality due to accident. Except AI-based cars, various applications of AI-based vehicle are as follows:

1. Automated trucking
2. Intelligent transportation systems
3. Optimization of route and traffic flow
4. Intelligent railway system through AI

3.4 Role of AI in Decision-Making

In the present era of Artificial Intelligence and Big data analytics, disagree the essential of a new machine and human symbiosis and require more analysis of "how machines need to work to expand the capabilities of human". "Wilson and Daugherty [24] suggested that companies that install AI mainly to displace man power will see only short-term production growth". If such things are correct, why and how industries will use AI for substituting worker not bring the long-term growth and what is the process to eliminate this inadequacy [22, 23].

"Wilson and Daugherty" [24] as well stated that organizations be able to make profit from optimizing "collaboration between humans and artificial intelligence" and build up workers' "fusion skills" that allow them to work efficiently at the human machine model. Although, some AI methods do not have the potential to make clear the "reasoning process of decision making" similar to, how to resolve the Black-box problem, i.e., "knowing why judgments are made in a certain way and offer justification to AI users", to solve this problem, Miller [22] examines that there has been a current resurrection in the field of explainable AI due to researchers want to make AI applications more comprehensible [24].

3.4.1 *Strategic Management and AI*

"Strategic Management" is the analysis, planning, controlling, and evaluation of all the department of organization to attain its target and objectives. The strategic management technique can be illustrated by following steps:

- Strategic intent
- Strategy formulation
- Strategy implementation
- Strategic evaluation

Strategic Management Planning Process

- Gather inputs
- SWOT analysis
- Review inputs
- Strategic matrix
- Define strategies
- Final review

Process of strategy management with AI in business is useful in marketing decision-making, customer relationship management, recommendation, social relation etc.

3.4.2 *Key Challenges to Adopt AI in India*

"The preceding analysis of focus sectors like Healthcare, Agriculture, Education, Smart Cities and Infrastructure, and Smart Mobility and Transport, highlight the potential of AI tools and technologies in transforming the sectors and state of Indian economy as a whole. Adopting a narrow view and focusing on the challenges for a specific sector, the barriers to developing a robust set of AI applications may seem contextual and limited to that sector" [25]. For example, healthcare sector adopts the following factors:

1. Without joint effort between a variety of stakeholders: "While India has agree to electronic health record (EHR) policy, distribution of data among various hospital chains still remains in progress, since various hospital chains have agreed to different explanation of digitizing records" [25].
2. Appropriate data is not available and there is unavailability of robust "open clinical data sets."

3. "Concerns on privacy and security of data, including lack of formal regulation around anonymization of data" [25].

Nevertheless, during investigations across the spotlight sectors, the challenges are determined based on the following common factors:

1. Unavailability of enabled data ecosystems
2. Less concentration of AI innovation
3. Resource price is high and low knowledge for accepting AI in business model
4. Insufficient availability of AI-proficient and skilled professionals
5. Undecided ethical, security and privacy policies

Distasteful "Intellectual Property Rights (IPR)" to incentivize research and acceptance of AI.

3.5 Conclusion

Artificial intelligence is the most dynamic technology which can make a great impact on industries and society. Super intelligence of machine, unlimited power of computing, declination of storage data cost give the detailed information about the technology. Survey and case studies completed in India and for India also give the updated scenarios of technology, opportunities, capital augmentation, and AI global market situation. Potential of AI in health sector, agriculture, education, transportation are also described in detail. At last, this chapter also describes the role of AI in decision-making, strategic management and key challenges for the adoption of AI technology in industries and market.

References

1. S. Stacey et al., *Worldwide Semi-Annual Cognitive Artificial Intelligence Systems Spending Guide* (IDC, Needham, 2017)
2. A. Khatri, T. Choudhury, T. P. Singh, M. Shamoon, Video analytics based identification and tracking in smart spaces, in 2019 International Conference on Contemporary Computing and Informatics (IC3I), 2019, pp. 261–267
3. A. Gupta, A. Singhal, A.S. Sabitha, T. Choudhury, Water utility management using Data Analytics and Geospatial Analytics for state of Delhi, in Proceedings of the 2nd International Conference on Green Computing and Internet of Things, ICGCIoT, 2018. https://doi.org/10.1109/ICGCIoT.2018.8752970
4. T. Choudhury, A.S. Chhabra, P. Kumar, S. Sharma, A recent trends on big data analytics, in 2016 International Conference System Modeling & Advancement in Research Trends (SMART), 2016, pp. 225–231
5. A. Kant, National strategy for artificial intelligence, Niti Aayog, 2018. https://niti.gov.in/writereaddata/files/document_publication/NationalStrategy-for-AI-Discussion-Paper.pdf
6. N.P. Jouppi et al., Datacenter performance analysis of a tensor processing unit, in ISCA '17: Proceedings of the 44th Annual International Symposium on Computer Architecture, 2017, pp.

1–12
7. S. Gupta, *Future of Jobs in India: A 2022 Perspective* (FICCI, New Delhi, 2017) http://ficci.in/study-page.asp?spid=22951&sectorid=74
8. R.M. Menon, M. Vazirani, P. Roy, *Rewire for Growth: Accelerating India's Economic Growth with Artificial Intelligence* (Accenture, Dublin, 2018) https://www.accenture.com/_acnmedia/PDF-68/Accenture-ReWire-For-Growth-POV-19-12-Final.pdf#zoom=50
9. P.K. Mittal, *Indian Healthcare Start-Ups—An Inside Look into Funding* (FICCI, New Delhi, 2016) http://ficci.in/study-page.asp?spid=20767&sectorid=18
10. S. Herbst, A. Plasschaert, *PwC Analysis* (PWC, London) https00://www.pwc.com/gx/en/research-insights.html
11. National Health Policy 2015 Draft, Ministry of Health & Family Welfare December, 2014. https://www.nhp.gov.in/sites/default/files/pdf/draft_national_health_policy_2015.pdf
12. Economic Survey 2018, Union Budget 2018. https://www.indiabudget.gov.in/
13. Financial Express Bureau, Doubling farmers' income: Dalwai committee empowered to oversee implementation, Financial Express, 4 Jan, 2019, https://www.financialexpress.com/market/doubling-farmers-income-dalwai-committee-empowered-to-oversee-implementation/1432897
14. R. Chand, R. Saxena, S. Rana, Estimates and analysis of farm income in India, 2015. Economics & Political Weekly, https://www.epw.in/journal/2015/22/special-articles/estimates-and-analysis-farm-income-india-1983-84-2011-12.html
15. Artificial Intelligence (Ai) in medical diagnostics market by component (software, service), application (in vivo, radiology, OBGY, MRI, CT, Ultrasound, IVD), end user (hospital, diagnostic laboratory, diagnostic imaging center)—Global Forecast to 2025. Markets and Markets, https://www.marketsandmarkets.com/
16. S.D. D'Cunha, For India's farmers it's agtech startups, not government, that is key, Forbes.com. https://www.forbes.com/sites/suparnadutt/2018/01/08/for-indias-farmers-its-agtech-startups-not-government-that-is-key/
17. Buckner Bill, Why FoodTech & AgTech? Agfunder.com, https://agfunder.com/about-us/mentors/
18. DFI Committee Report, Department of Agriculture, Cooperation & Farmers Welfare, http://agricoop.nic.in/
19. Ed Tech Adoption Survey, Cross Square Foundation, https://www.centralsquarefoundation.org/edtech/
20. Economic Survey, Union Budget 2017, https://www.indiabudget.gov.in/budget2017-2018/survey.asp
21. R. Mohan, et al., *India Transport Report: Moving India to 2032*, NTDPC, http://www.aitd.net.in/national-transportd-evelopment-policy-committee-report.php
22. S. Miller, AI: Augmentation, more so than automation. Asian Manage. Insight. **5**(1), 1–20 (2018a)
23. M.H. Jarrahi, Artificial intelligence and the future of work: Human-AI symbiosis in organizational decision making. Bus. Horiz. **61**(4), 577–586 (2018). https://doi.org/10.1016/j.bushor.2018.03.007
24. J. Wilson, P.R. Daugherty, COLLABORATIVE INTELLIGENCE humans and Al are joining forces. Harv. Bus. Rev. **96**(4), 115–123 (2018)
25. Shaping the future of technology governance: Artificial intelligence and machine learning. *World Economic Forum,*https://www.weforum.org/

Chapter 4
Artificial Intelligence: Game Changer in Management Strategies

Om Prakash Gusai and Ankur Rani

4.1 Introduction

In today's world, due to more advancement in "technology of deep machine learning" different fields of the economy can be seen as a huge global automation. Most of the known standard task like projects related to e-commerce for getting more efficiency, task related to production line in various industries is managed by intelligent machines. This global change leads to advancement in every field. It can become a reasonable factor about the human beings' replacement in different job roles and verticals of industries. Artificial intelligence just not only leads to adding value for providing great opportunities for management purpose but also poses to help in managers' complex tasks related to the organization. In today's competition, organizations have to create that type of structure related to training and strategy with adoption of use of artificial intelligence which helps to focus on those particular tasks which consists of creativity and judgment-related skills. Managers have to revise the traditional principles of management with full of cooperation among human beings and deep machine learning [1].

O. P. Gusai
Department of Commerce, Motilal Nehru College, University of Delhi, New Delhi, India
e-mail: opgusai@mln.du.ac.in

A. Rani (✉)
Department of Commerce, Baba Mastnath University, Rohtak, India

P. M. Jeyanthi et al. (eds.), *Decision Intelligence Analytics and the Implementation of Strategic Business Management*, EAI/Springer Innovations in Communication and Computing, https://doi.org/10.1007/978-3-030-82763-2_4

4.2 Concept of Artificial Intelligence

After examining the concept of artificial intelligence from the literature, the term can be defined as a structure of full process which consists of critical thinking, typical ideas of reasoning, getting the outcome by using particular thought of information and comprehending objects and its realities. The concept of "artificial intelligence" has the origin of "Latin culture," and it is called "intelligere" in Latin language. This concept is based on the binary values. When the word breaks in two sub-words, it generates inters and lingerie. The word inters means between and lingerie means to select or to read. Basically "intelligere" is created by dual structure. The term "Artificial Intelligence" is comprehensive in its nature and logical to some extent. Artificial intelligence creates structure which performs various functions related to structure of non-organic forms. The literature finds that the definition of artificial intelligence is generally not accepted. According to the "barr and feigenbaum," AI is the intelligent form of structure that is designed from computer science. It has the ability to solve various types of problems, learning qualifications, understanding the various patterns in computer language. This is artificial intelligence [2]. According to Nilsson, artificial intelligence creates a model which works on the human brain with the logic and make in every aspect of science field. Artificial intelligence and its applications performs well and provide a better understanding for making strategic decisions [3].

4.3 Background of "AI"

The idea about artificial intelligence was first identified by the practical test created by Alan Turing in 1950. The test was based on artificial intelligence and its various applications. "Turing," in 1950, assumed that for judging artificial intelligence, several arguments are attached to it. The Turing test performs various tasks like if the person observes things and his observance is not able to make a difference between the conversation with the particular machine and the conversation with the particular human being, then according to this test the machine is an intelligent element [4]. Earlier the focus of applications were to develop those computers programs which can be competed with the humans' mind for example the structural games like a game of Chess. When the "Dartmouth conference" was held in 1956, John McCarthy coined artificial intelligence, and the conference was the birthmark of artificial intelligence in a discipline scientifically [5]. The Turing test was coined by ISM department in 1940s, different scientists discuss about the machine which can think beyond its extent and that it can perform various tasks that human being generally performs. After few months, students of "Princeton" built the expert "neural networks"; it was first development by "artificial university researchers." They used "300 vacuum tubes" for creating the software which has surplus "gyro pilot." This database was used for knowledge experts, and finally it was discovered in 1955. The Carnegie Institute of Technology which is now

Carnegie Mellon is another artificial intelligence university which discovered the first program of artificial intelligence, theories of logic. In the twenty-first century, artificial intelligence boomed in every field, "marvin lee" advances in the field of deep learning which was underpinned with "minsky" founded at the laboratory of artificial intelligence. Others also worked on semantic analytics for various AI operations [6].

4.4 The Digital Business Vagueness

Today is digitalization's era which is full of uncertainty where there are rules of game changers all over again and again. If today someone follows to learn the technology, the next day the other technology generates and dominates the previous one. The conditions of a particular organizations' environment which are not in the routine format create the level of uncertainty. The organization must be protective against the conditions which can be unpredictable. Today for managing the uncertainty level, various organizations perform managerial task to control it.

Mission is the new way for engagement with the customers and for performing tasks in doing business. Every organization has its own perspective because most of the organizations do not know exactly from where to start to cope up with the new era of digitalization, robotics sector, and ultimately with artificial intelligence [7].

4.5 Digital Transformation in Management

The starting point for the creation of the strategies is vision. Digital development affects the enterprise, its works, and its dynamics. This is the new technology which the organization tries to adopt the concepts of machine learning, automation, internet of things, robotics, and so on. For the stakeholders, it can have a different meaning. It can be developed with the new vision of an organization; obviously the change will be there. This form of transformation creates different perspective of stakeholders regarding the receiving of particular data from the computer system. Viewpoint provides the platform where the elements connected to the responsibility define the position of the stakeholders. The leader should direct the stakeholder's viewpoint for working with harmony; it creates the value which will be delivered by means of vision [8]. The various organizations have adopted the applications of new digital technology which has changed the way of processes and matter of earnings. There are five phases which define the digitalization era. These can be "ambition," "design," "scale," "delivery," and "refine" [9].

When enterprise is going to have an interest and excitement towards the adoption of digitalized way of business, it is called the ambition. But when a new innovativeness and creativity has been added to creating and producing the products and services comes under the term "Design". The deliver phase includes the delivery of the viable products. Scaling is the measurement of business and its operations fully. To refine the task is the phase with digital business and the opportunity related to be optimized. It can have multiple difficulties because of the transformation of management in a new form. That should understand the transformation as it is a slow process, the personnel should execute the plan in a proper format. Various organizations and companies should execute the five phases of digitalization transformation in a new era. The five phases are as follows.

4.5.1 Ambition of the Organization

To achieve an ambition of the organization, there should be an adoption of some necessary steps of innovation like brainstorming session of the particular business units' performers. The organizations should have the acquired funds for performing the tasks in continuation [10].

4.5.2 To Design a Product

With the use of digital transformation, the organizations' management should have a plan to design that type of product which will be viable and which will receive a lot of attention because of the brainstorming and the ideas which are innovative. There are various different channels which will be used for serving the product. The vision of the viable product must Omni channelized. All the parts of the organizations must work in holistic manner. To establishing the designed product with the end users, it is advisable to take the viewpoints of the end users and tested their experiences. It cannot be ignored that the most important factor to design a product is ultimately the end users. The Understanding of the peoples' behavior, should be the main purpose of the organization while designing for the transformation phase.

4.5.3 Deliver Phase

When the product has been designed, the next phase will be how to deliver it. In the real world, the organization should have the target of 3 months for delivering. There should be pilot testing in the production line. The development in the capability to desire the risk factor should also be examined.

4.5.4 Scaling

It is the measurement phase where the organization can track or to understand the designs success. The scaling have multiple functions which can be performed on the digital platforms, criteria to be followed by the organization to fill up the necessary moves and for achieving the mission of the organization.

4.5.5 To Refine

This phase comes in the digital transformation when the product is launched and the enterprise is at the pace of learning the new digital capabilities. The revenues increment, and if the organization is on the move of success, the refinement and the mission are completed with the feedback of customers.

4.6 Artificial Intelligence in Organization

According to Avolio, the e-leadership mediated things due for a change in organization. The literature reviewed that it is the combination of science and leadership practices which force the organization to adopt the advancing technology. The argument was focusing on the prediction of desirable practices. Artificial intelligence for managers replaces or enhances the task related to their work. It can affect indirectly on the basis of level like level of society, the organizational structure, level of task manager [11].

4.7 Who Is Manager?

A manager is an individual who is in a supervisory position from the c-suite executive level to the factory floor level. The c-suite level of the top manager is a member of management, for example, the CEO. The middle manager can be a plant manager or manager of divisions or a senior administrative work manager who performs various operational tasks. The frontline managers are those managers who are an office manager or manager of any department, crew leaders, and team leader who perform the smaller tasks or smaller projects.

4.8 "Strategic Management"

It is the management of planning, interpretation, controlling facts, and assessing all particular things which lead to the organization towards for achieving the required goals strategically. Strategic management requires various steps that start with the organizations' mission, when the organization has idea about what the organization really wants to establish, and next would be vision which the organizations' management have clear thought about what the organization wants to be in the future, what really it wants to acquire [12].

Basically it is a situation analysis which is going to be used to achieve the vision. The analysis also considers the organization management tasks and other questions arises like what is the organization strength and weaknesses, what are its opportunities and threats, how it can be handled in a proper manner, and after the analysis of strategic planning which occurs for achieving the particular objectives.

The artificial intelligence is not only the simple technology which is used in the business, but it is a system of programs that are able to control the whole organization. It is a connection between the people, the processes, and the strategy [13].

4.8.1 Connection Between Strategic Management and Artificial Intelligence

All the works related to data collection, to get information, the analysis of data are the basis for strategic management for evaluating the outcomes. On the basis of this information and the indicators, the particular organization creates a strategy for designing the better outcomes. The organization can prepare the full plan from starting to the end point like from where it wants to start, what the organization really wants in the future, what to achieve. In the advancement of artificial intelligence, strategic management can be more fruitful. Artificial intelligence persuades top management, making the strategic decisions for achieving the goals which are realistic in the line of mission and vision of the organization. It can also create the great value which will be analyzed for futuristic protectiveness, providing better leaders to the market [14].

4.8.2 Improvement and Redefining in the Organization Strategies Vis-a-Vis AI

As artificial intelligence is the part of computer science which helps to improve the designing of strategic planning and helps to improve the various level of strategy like vision, mission of the organization, the content creation, and objectives. Formulation of strategy is the typical task for the manager because it involves all the internal and external factors, situation analysis, and performing those task with suitable for establishing the strategic plan. AI helps to mold the wrong path to the right path with the evaluation, monitoring in the implementation phase. Each step in the strategic planning process required artificial intelligence for performing in better aspects. Artificial intelligence connects internal and external resources. It works like elastic between the process, competitors, and the resources. The artificial intelligence collect the facts and analyses performance of the organizations regarding its strategic plans and its implementation till the interpretation. On the basis of KIPS, artificial intelligence has the ability to change the organization strategy. It redefines and monitors the strategic models and various theories [15].

4.8.3 Artificial Intelligence on Strength, Weakness, Opportunities, Threat (SWOT)

When the organization follows the situation analysis with the use of artificial intelligence, it can be more accurate and detailed. Strengths and weaknesses are the parts of internal analysis of the organization, and opportunities and threats are the part of external analysis of organization. The capability is in the personnel, the recruitment process, the situation of finance and marketing, and other competitive advantages come under the strengths. The weaknesses can be like problems in cash flow statements, fill the gap between competitive strength, to manage the supply chain, and so on. Artificial intelligence provides the intelligence on how to avoid the threats and to convert the full energy in opportunities which can become the strength of an organization [16].

4.9 Conclusion

The role of artificial intelligence is going to be strengthening every field, business management, and external environment of the organization; all are affected by AI. The use of artificial intelligence in strategic planning and situation analysis is going to be a targeted purpose as it plays the role of game changer in the strategic management; it monitors all levels of business, all steps in the strategic planning process. It is just not the tool which is used by managers only but somehow it is like a companion for effective performance in strategic management.

References

1. V. Chernova, Artificial intelligence in management: Challenges, in 38th International Scientific Conference on Economic and Social Development, 2019, March, 21–22
2. J. McGregor, A. Barr, E. Feigenbaum, The handbook of artificial intelligence, volume 1. Math. Gaz. **66**(438), 336 (1982). https://doi.org/10.2307/3615550
3. N. Nilsson, *The Quest for Artificial Intelligence: A History of Ideas and Achievements* (Cambridge University Press, Cambridge, 2009)
4. A. Turing, Computing machinery and intelligence. Mind **9**(236), 433–460 (1950). https://doi.org/10.1093/mind/lix.236.433
5. J. McCarthy, M.L. Minsky, N. Rochester, C.E. Shannon, A proposal for the Dartmouth summer research project on artificial intelligence, 2006, August 31, 1955
6. L. Gugerty, Newell and Simon's logic theorist: Historical background and impact on cognitive modeling. Proc. Human Factors Ergon. Soc. Ann. Meet **50**(9), 880–884 (2006). https://doi.org/10.1177/154193120605000904
7. K. Dorner, D. Edelman, What 'digital' really means, 2015. https://www.mckinsey.com/industries/high-tech/our-in sights/what-digital-really-means. Accessed 15 Nov 2020
8. İ. Met, D. Kabukçu, G. Uzunoğulları, Ü. Soyalp, T. Dakdevir, Transformation of business model in finance sector with artificial intelligence and robotic process automation, in *Contri-*

butions to Management Science, (Springer, Cham, 2020). https://doi.org/10.1007/978-3-030-29739-8_1
9. D. Scheibenreif, C. Geschickter, A CIO's guide to gartner's digital business research, 2018. https://www.gartner.com/document/3862963. Accessed 12 Nov 2020
10. V. D'Orazio, P. Redshaw, 2019 CIO agenda: banking and investment services industry insights, 2018. https://www.gartner.com. Accessed 15 Nov, 2020
11. B.H. Rashmi, Impact of artificial intelligence on cyber security. Int. J. Comp. Sci. Eng. **6**(12), 341–343 (2018). https://doi.org/10.26438/ijcse/v6i12.341343
12. M. Sundu, S. Ozdemir, Effect Artif. Intell. Manage. Process **2**(1), 22–41 (2020). https://doi.org/10.4018/978-1-7998-2577-7.ch003
13. E. Tsui, B. Garner, S. Staab, The role of artificial intelligence in knowledge management. Knowl.-Based Syst. **13**(5), 235–239 (2000). https://doi.org/10.1016/s0950-7051(00)00093-9
14. C. Holloway, Strategic management and artificial intelligence. Long Range Plann. **16**(5), 89–93 (1983). https://doi.org/10.1016/0024-6301(83)90082-1
15. W.E. Spangler, The role of artificial intelligence in understanding the strategic decision-making process. IEEE Trans. Knowl. Data Eng. **3**(2), 149–159 (1991)
16. R. Orwig, H. Chen, D. Vogel, J.F. Nunamaker, A multems and artificial intelligence. Group Decis. Negot. **6**(1), 37–59 (1997)

Chapter 5
Prospects and Future of Artificial Intelligence (AI) in Business Strategies

Shekhar Chandra Joshi and Yougal Joshi

5.1 Introduction

Intelligence helps each other to work and control in several ways of life. Understanding and appreciation both can do the same control. Italian rationalist Niccolo Machiavelli had told three sorts of knowledge: first, comprehends things for itself; second, what others can comprehend; and third, comprehends neither for itself nor through others ([1]). It is true that no one would like to be placed in the third category that depends on his/her intelligence.

Artificial is the product of manmade instead of growing naturally. It might be a copy of something natural having human behavior too. It is a fact that human and its behavioral activity is full of art and communication. Through the ages, the civilization human intelligence ascribed in each walk of life for picking up, thinking, critical thinking, discernment, and utilizing language.

In a design's point of view- science, technology and innovation along with a blend of craftsmanship, and the assistance of new technologies with the help of Artificial Intelligence (AI) and Software Engineering to perform over the machine is act like human. Hence, that is called a humanoid. Although, the intelligence they have acquired is being developed by scientists and technicians and related people.

Robots have been working with intelligence and developed hence providing citizenship to the Remote "Sophia" by Saudi Arabia that became the first country of its kind in the world in 2017. This remote was created by Hanson robotics. The simple robots can move forward, backward, move left, move right, hold, pick, and

S. C. Joshi (✉)
Soban Singh Jeena, (Campus) University, Almora, India

Y. Joshi
Kumaun University, Nainital, India

P. M. Jeyanthi et al. (eds.), *Decision Intelligence Analytics and the Implementation of Strategic Business Management*, EAI/Springer Innovations in Communication and Computing, https://doi.org/10.1007/978-3-030-82763-2_5

drop the things up to the marked instructions and directions. Different subtleties will be featured ahead. Japanese with the assistance of AI are distinguishing the indications of inebriation in travelers at train stations. PCs and PC programming have assumed a significant part in it. Stuart et al. [2] have unveiled numerous things toward this path.

5.2 Literature Review Including Research Gap

The artificial intelligence (AI) is a modern approach to create a world of humanoid robots as a needful workforce. The wish of technicians and scientists as well as businessmen increased toward the invention and production of AI. The needful demand and supply of AI products gave a vast vision to business strategies. Instead of published literature, online material is available that needs to be researched and compiled. Hence, in this chapter as a development toward the new branch as well as promoting the craze of online/digital learning and research, the maximum references used (quoted at the end) have been based online. As an artist, preparing this chapter for me is a challenge while Dr. Yugal Joshi is acquainted in the scientific field.

5.3 Materials and Methods

Materials and methods related to AI have been taken from the worldwide companies and industries for logistics, automotive, industrial design, and industrial and consumer robotics base. Industries for education, gaming industry; companies and industries for health care; academia, defense, aerospace, energy, hardware and software; etc. have been sampled. The detailed discussion tabled separately is described in the following.

5.3.1 Worldwide Companies or Owners of Humanoid and Related to AI

Technicians and others related to man-made brainpower (AI) work easily to advance the discourse acknowledgment, machine interpretation, self-governing vehicles, and family mechanical technology. Thinking, AI, and PC vision just as algorithmic milestones assume a significant part in it. Art and artistic engineering give them the final shape to the products. The root of AI is robotic arm doing a lot of work. The use of robotic arms can be seen in the factories producing daily use items. These robotic arms and robots as well as humanoid robots can be seen in the factories, industries, health care, education, and defense. These are prepared by the companies or owners of the humanoid [7, 9]. The names of companies or owners of robot, robotics, and humanoid with their work force are shown in the Tables 5.1, 5.2, 5.3, 5.4, 5.5, 5.6, 5.7, and 5.8.

5.3.2 *Workforce Domain Contains Retention*

It seems AI has adequate potential in the field as indicated in Sect. 5.3.1. The worldwide companies or owners of humanoid and related to AI have been developing their business.

Japanese organizations with "3E-A18," "3E-B18," "3E-C18," "3E-D18" bots are intended to help individuals by and large, from cataclysmic events to entertainment. NASA PUFFER robots have the quality to scroll on uneven surfaces.

Nouvelle man-made consciousness, a way to deal with computerized reasoning (AI) spearheaded at the Massachusetts Institute of Technology (MIT) AI Laboratory

Table 5.1 AI-related companies and industries for logistics, automotive, industrial design, and industrial and consumer robotics base

Working area for	Name of company/industry	Location of company/industry
Logistics base makes an autonomous robotic cart	Canvas Technology Company	Boulder, Colorado
Logistics, computer vision for people and goods	Piaggio Fast Forward	Boston, Massachusetts
Industrial and consumer robotics applications	Energid Technologies	Cambridge, Massachusetts
Automotive, Rethink's collaborative robots (cobots)	Rethink Robotics	Boston, Massachusetts
Assembling, warehousing and satisfaction make self-ruling robots transport, lift	Vecna Robotics	Cambridge, Massachusetts
Engineering and robotics design	Boston Dynamics	Waltham, Massachusetts

Table 5.2 AI-related companies and industries for education, gaming industry

Working area for	Name of company/industry	Location of company/industry
Education and gaming industry	Sphero	Boulder, Colorado
Education and health care	Anybots	San Jose, California
Makes an automated framework—Called cortex, which can be utilized in an assortment of conditions	Amp Robotics	Denver, Colorado
Computational reasoning and help improve kids issue solvers in interconnected conditions through play-based learning	Modular Robotics	Boulder, Colorado

Table 5.3 AI-related companies and industries for health care

Working area for	Name of company/industry	Location of company/industry
People in everyday environments	Diligent Robotics	Austin, Texas
Medical services and schooling. Furnished with a speaker, camer.a and video screen	Anybots	San Jose, California
Arms and hands/progressed automated controllers for an assortment of utilizations	Barrett Technology	Newton, Massachusetts
Mechanical, farming, and medical care. Energid's acting SDK programming empowers progressed constant movement control	Energid Technologies	Cambridge, Massachusetts
Medical services and its intuitive robots are utilized for negligibly obtrusive medical procedure	Intuitive Surgical	Sunnyvale, California
Consumer, entertainment, service, healthcare, and research applications	Hanson Robotics	Hong Kong

Table 5.4 AI-related companies and industries for machine learning and advanced companion robots and maker of mechanical technology programming and equipment

Working area for	Name of company/industry	Location of company/industry
Movement arranging, progressed converse kinematics, constant control, crash evasion, 2D route, augmented reality, robot demonstrating, and so forth	Picknik Robotics	Boulder, Colorado
Buddy robots; upgrade singular health and personal satisfaction	Embodied Robotics	Pasadena, California
Robotics as a maker of mechanical technology programming and equipment, advance the best in class in self-ruling advanced mechanics	Willow Garage	Palo Alto, California
Consumer, entertainment, service, healthcare, and research applications	Hanson Robotics	Hong Kong

by the Australian American researcher Rodney Brooks during the last 50% of the 1980s. Specialists of nouvelle AI affirm that genuine knowledge includes the capacity to work in a certifiable climate. One renowned illustration of nouvelle AI is Brooks' robot Herbert (named after the AI pioneer Herbert Simon), whose climate is the bustling workplaces of the MIT AI Laboratory. All the more as of late, Brooks built models of versatile robots for investigating the outside of Mars [8, 10].

Table 5.5 AI-related companies and industries for academia, defense, aerospace, energy, hardware, and software

Working area for	Name of company/industry	Location of company/industry
Different robots that have human- and animal-like dexterity	Boston Dynamics	Waltham, Massachusetts
Aviation, energy, and equipment; it assembles three various types of robots that perform inconceivably various capacities	Sarcos	Seattle, Washington
General dynamics, Bluefin makes unmanned and autonomous underwater vehicles (UUV/AUV)	Bluefin Robotics	Quincy, Massachusetts,
Horticulture, protection, and man-made consciousness	Applied	Austin, Texas.
Academia, defense, and aerospace	Honeybee Robotics	Brooklyn, New York
During disaster and for NASA	Honda e2-dr [14]	Japan

Table 5.6 AI-related companies and industries for pet, agriculture, industrial, public, and public safety and outdoor task

Working area for	Name of company/industry	Location of company/industry
Making an automated cat toy called Mousr	Petronics	Chicago, Illinois
Making robotic Aibo dog [11] and other animal	Sony Aibo	Tokyo Japan
Agriculture, defense, and artificial intelligence for (UAVs)	Applied Aeronautics	Austin, Texas.
Organization's self-driving SnowBot pro. Distantly controlled on the web	Left Hand Robotics	Longmont, Colorado
Public wellbeing applications, especially those including firemen and police	Drone Sense	Austin, Texas
Modern, farming, and medical care. Energid's actin SDK programming empowers	Energid Technologies	Cambridge, Massachusetts
Agribusiness, HV-100 model was known the world's first completely independent robot	Harvest Automation Industry	Billerica, Massachusetts
Making an assortment of brilliant vacuuming, floor-scouring, and cleaning gadgets	Irobot Robotics	Bedford, Massachusetts
Those to help (controlled support) who've encountered loss of motion or debilitating in their grasp and arms	Myomo	Cambridge, Massachusetts

Table 5.7 AI-related companies in Japan

Working area for	Name of company/industry	Location of company/industry
Automated food robots	Connected Robotics	Tokyo
Man-made intelligence and profound learning-centered startup, as of late revealed the world's first robot fit for cleaning up a room	Preferred Networks Inc.	Tokyo
Progressed, intuitive robot to go about as a social ally for the old, explicitly intended to animate patients with dementia, Alzheimer's, and other intellectual issues	Paro Therapeutic	Nanto, Toyama
Creates distant controlled robots that can reflect one's activities, regardless of whether on the contrary side of the globe	TX Inc. (Telexistence)	Tokyo
Produced cube a "portable, personal assistant robot"	Plen Robotics	Osaka
Gives AI-based programming to advanced mechanics and self-governing vehicles	Ascent Robotics	Tokyo
Advanced mechanics on selling exploratory vehicles with self-governing driving innovation introduced as stages for innovative work	ZMP Inc	Tokyo
Companies automate manufacturing and warehousing processes; "cooperative working robots"—automated machines	Life Robotics	Koto City, Tokyo
Simulated intelligence-based Mujin's regulators permit robots to "think" through their developments with no pre-programming, acclimating to a circumstance as it happens	Mujin Inc.	Koto-ku, Tokyo

Stanford college specialists have fostered a snake-like [13] robot that develops like a plant by getting through difficult to arrive at places in a debacle or other crisis tasks. The snake robot can turn and bend round troublesome corners, contingent upon what it sees from the camera.

Like the tentacle designing an octopus and its grip, the octopus gripper robotic arm of Festo holds smooth contact help in the smooth working for an industrial environment. There is Google-owned robotics firm Boston dynamics "handle" that is designed for heavy lifting and carrying loads of up to 50 kg. The Bank Mizuho with IBM and Softbank using humanoid robots to provide customer support. An

Table 5.8 AI-related companies and their workforce in India

Working area for	Name of company/industry	Location of company/industry
Artificial intelligence, block chain, RPA, and so forth and supplementing it with human-driven UI/UX plan, information	Day one technologies	Bengaluru
Big data related to intelligence services and machine learning services	Sigma data Systems	Bengaluru
Versatile/app/AR/VR/AI/game development	Quytech	Gurgaon
Redone AI: Machine power + human accuracy	SetuServ [17]	Hyderabad
Block chain, RPA, artificial intelligence and ML, IoT, business intelligence [19]	Prolitus Technologies	Noida
Calculation experts, data scientists, machine learning specialists, and technology consultants	Consagous Technologies	Bengaluru
Leading artificial intelligence, mobility, IoT, cloud solution provider	Fusion Informatics Limited	Bengaluru
AI & ML services simplified	Cognitive Machines	Bengaluru
Artificial intelligence services, machine learning development, digital marketing services, and many more	Ecosmob Technologies Pvt. Ltd.	Ahmedabad
IT, block chain, AI dataset, and annotation	INCN Technology	Gurgaon

Indus International School in Bengaluru, India has started teaching with the help of humanoid major science subjects.

Above all and many others are the examples that create a workforce domain containing retention in the field for the future. Honda and its Asimo are working on a valued task by presenting model fiasco alleviation robot fit for exploring through perilous, complex conditions.

5.4 Case Study and Application

Case study and application always help in the research. These are described in the following.

5.4.1 Workforce Planning

Countries engaged in AI works have built their strength in the field defeating others. America, China, Japan, Europe, and India have their own workforce planning regarding AI. This could be understood easily in the following.

An American scholastic, Thomas Davenport [5], stated that businesses believe AI ought to be strong instead of hot or splashy. Exhorting, he portrayed the significant AI advances, clarified how they are being utilized, and showed an account of the AI work done by huge business undertakings like Amazon and Google [6].

China devotes significant efforts to realize its goal of becoming a leading global player in artificial intelligence (AI) through its workforce planning. The Chinese companies' contribution in AI across the country as Chinese contribution in AI is shown in the following Fig. 5.1.

Japan is additionally giving more confirmations and awards from the public authority to Japanese colleges for by and large turn of events and market of AI including understudies and laborers. The Tokyo-based firm, which is worth generally $2 billion, as indicated by CB Insights, is an image of Japan's broad vital development drive, where AI and mechanical technology are seen as keys to both addressing social issues and accomplishing new financial development [16].

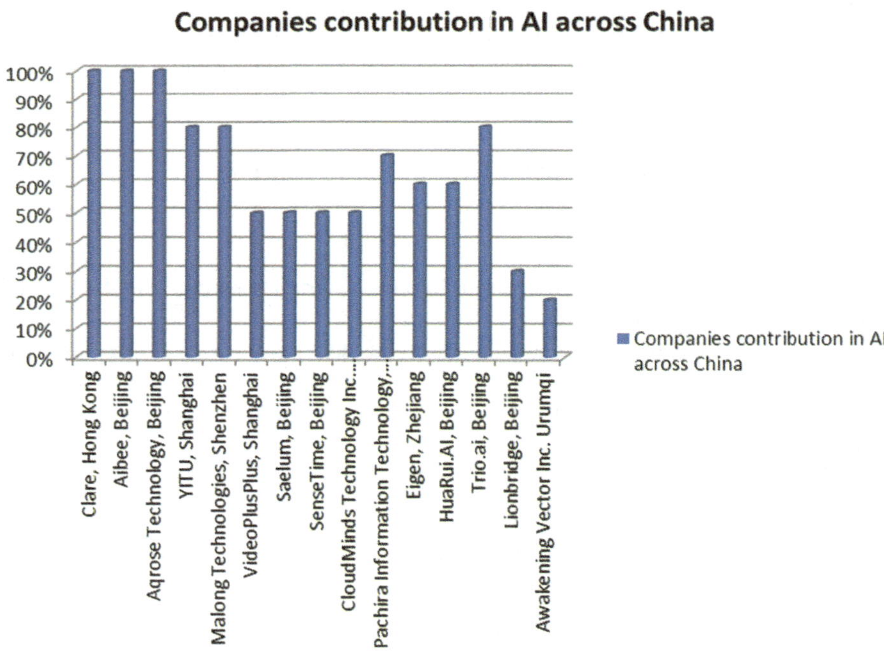

Fig. 5.1 Chinese contribution in AI

In the AI market, Japanese organizations are setting up AI R&D bases focused in the US, Toyota, and Hitachi to run fast in the field. Notwithstanding applications in self-driving vehicles, Toyota is intending to utilize AI to improve individuals' lives through robots.

As per IFR research, robot deals in India expanded by 27% to another pinnacle of 2627 units in India—practically equivalent to that in Thailand. South Korea with the assistance of AI Robot "Aglio Kim" is serving in the café. They did the best utilization of Robot in COVID-19 [20] for checking the temperature of representatives and furthermore administer hand sanitizer and clean the floor.

In 2019, South Korea had introduced 855 modern robots for each 10,000 workers [18]. That is basically because of the progress with establishment of high volume robots in the hardware and electric areas. Germany and Japan are prestigious for their car ventures and they have thickness levels of just around 350 for each 10,000 laborers.

Europe and other EU individuals have set out seven key prerequisites that AI frameworks should meet to be dependable:

5.4.1.1 Human Office and Oversight

Artificial intelligence structures should empower people, allowing them to make instructed decisions and developing their focal rights. All the while, fitting oversight frameworks ought to be ensured, which can be cultivated through human-on top of it, human-on-the-circle, and human-all together moves close.

5.4.1.2 Specialized Robustness and Well-being

Artificial intelligence systems ought to be intense and secure. They ought to be secured, ensuring a fall back game plan if something turns out seriously, similarly as being exact, trustworthy, and reproducible.

5.4.1.3 Security and Data Organization

Rather than ensuring full respect for assurance and data protection, acceptable data organization instruments, thinking about the quality and genuineness of the data, and ensuring legitimized induction to data.

5.4.1.4 Straightforwardness

The data, structure, and AI game plans should be clear. Conspicuousness frameworks can help achieving this.

5.4.1.5 Assortment, Non-isolation, and Sensibility

Unfair inclination ought to be avoided. Empowering assortment, AI systems should be available to all, paying little psyche to any incapacitate, and incorporate material accomplices all through their entire life circle.

5.4.1.6 Social and Common Success

AI systems should benefit each and every person, remembering individuals for what is to come. It ought to therefore be ensured that they are possible and innocuous to the environment.

5.4.1.7 Obligation

Mechanisms should be set up to ensure obligation and duty regarding AI systems and their outcomes [12, 15, 16].

5.4.2 The Interest of Denmark, Finland, France, and Germany in AI is Likewise Significant

Denmark's objective is to make Danish organizations the best at utilizing advanced advances. France has intended to change France into a worldwide innovator in AI examination, preparing, and industry. As indicated by a report, France will foster an open information strategy to drive the appropriation and use of AI in areas where France as of now has the potential for AI greatness, like medical care.

The German government needs to reinforce and grow German and European exploration in AI. The public authority centers on the exchange of exploration results to the private area and the making of AI applications.

In the following section the reasons for the aforesaid facts are explained in a better way to understand:

5.4.2.1 Occupation Misfortunes AI in the Work Environment Have Meanings of Mass Occupation Misfortunes

Adnan said: "This is driven by the lack of understanding of what AI is and isn't. AI will not replace human ingenuity or creativity anytime soon. AI is a tool; overestimating its potential or incorrect application can lead to reduced efficiency." In this way, the inverse might be valid. A Gartner report found that AI could produce up to 2.3 million positions by 2020. Nonetheless, this contrasts for every industry:

public areas, medical services, and training will see some work increment, while assembling and development can expect work misfortunes.

5.4.2.2 Expenses

High-tech like AI accompanies a powerful sticker price. Essential AI for an individual advancement can arrive at a monstrous US$300,000.

5.4.2.3 Absence of Mindfulness

AI is not totally blundering free. Simply a year ago, a driverless Uber vehicle hit a lady going across a street in Tempe, Arizona. One contention is that self-sufficient frameworks like driverless vehicles and robots are instructed by virtual preparing situations which do not coordinate with genuine conditions.

While there are still issues that should be worked out with regard to AI, these are rare. Critically, its commitments to laborer security and profitability are verifiable.

Now educationists have been exploring the idea and products available in AI. Schools, colleges, and universities are adopting AI. In recent years, Indus International Public School Bangalore has introduced humanoid robots to assist the teacher in different subjects.

There are various types of learning as applied to computerized reasoning. The least complex is learning by experimentation. More testing is the issue of executing what is called speculation. Speculation includes applying past experience to closely resembling new circumstances [8, 10].

5.5 Safety Measures

Safety is the most important aspect to anyone, and workplace safety must be a top priority. According to a report somewhere in the range of 2017 and 2018, 144 individuals were executed grinding away in the UK. The report unveils that altogether, around 1.4 million individuals experience the ill effects of business related ailments and wounds. In addition, laborer wounds sway organizations with working days lost and pay installments. It is essential that businesses should guarantee laborer security. They ought to give appropriate work environment conditions and a protected workplace.

5.5.1 Diminishes Human Error

Human variables play a gigantic, blunder, factor in work environment security, with weakness and stress promptly adding to mishaps. Around 40% of US laborers experience the ill effects of weariness. In this way, one significant advantage of AI is its powerlessness to get focused, drained, or unwell. As such, AI well-being can downsize human variables in the working environment.

5.5.2 Attempts of Hazardous Undertakings

Strictly talking, drones themselves are not AI. In any case, as innovation propels, drones are rapidly consolidating AI—permitting them to settle on choices and work themselves. Accordingly, AI well-being makes rooftop reviews less expensive, more productive, and in particular, more secure.

5.5.3 Track Worker Location and That is Only the Tip of the Iceberg

Industrial wearable essential capacities include the following specialist area:

- Overseeing crucial signs, for example, pulse, and circulatory strain
- Alerting to ecological dangers
- Issuing data to telecommuters
- Reducing possibility of musculoskeletal wounds
- Improving staff preparing

5.5.4 Screens Workplace Harassment

A glad laborer is a profitable specialist. Utilizing AI to recognize work environment provocation is anticipated to turn into the standard. Artificial intelligence can analyze and feature working environment badgering. An equivalent and different labor force brings endless business benefits.

5.5.5 Work Environment Automation

When you hear mechanization in the work environment, you consequently think of work misfortunes.

5.6 Recruitment

The enrollment is expanding step by step in the field. HR helps in selecting, representative help, worker improvement, and coaching. AI-fueled collaborators (Chatbots), HR tech specialists concur on for 2018, and it is the utilization of AI-controlled associates in different pieces of HR [23]. In such a manner, Mya, Olivia, Beamery, Textio, Paññã, and so on give assistance likewise.

As indicated by a report "The presence of computerized reasoning and advanced mechanics in industry is developing very quickly." By 2020, that expanded to 113 across the assembling area. Asia presently has a robot thickness of 118 units for each 10,000 laborers and that figure is 114 and 103 in Europe and the Americas, individually. China is one of the nations recording the most elevated development levels in mechanical mechanization however no place a robot thickness like South Korea has [21, 22].

According to the prediction of *Vishv Arthik Manch* AI will generate 13.3 crore employment till 2022 while IT Consulting Company Gartner claims 20 lakh services through AI will be generated till that date. Bindra [3] stated that he ratio of division of labor among machine and man is currently 30% and 70% which would be 52% and 48% in 2025, respectively. According to a report of Forbes, a machine learning engineer offered a handsome salary 8–8.50 lakh yearly in the initial stage in India while in countries like America this salary is more than the article related to ML and AI or IOT [4]. Covering in the top ten demanding jobs profile, India is listed on top ten. Learning with Google AI has been promoting it.

Authors and experts like Kai-Fu Lee, Hans Moravec, Ray Kurzweil believe that jobs through AI are more secure. The robots would overwhelm people sooner rather than later by providing secure jobs in the field of psychology, medical, health-services, research and engineering related to AI, fiction composing, teaching, criminal law, computer science and designing, science and management. As per Lee, China will squash Silicon Valley since it has more information, hates protection, and contends all the more heartlessly. Established in 2016, Tokyo startup Ascent Robotics has raised $17.9 million to foster astute arrangements in mechanical advanced mechanics and self-ruling vehicles. They will probably help make a more self-governing future [12, 15, 16].

5.7 Authors' Contribution

The author have described companions or owners of humanoid robots, and their needful workforce domain contain retention, workforce planning, safety, and recruiting. They also attempt to explain how and in which places these humanoid robots are helpful across the world including India and up to which stage. This chapter reveals about workforce planning, safety, and recruiting. The inclusion of the present chapter in the book will strengthen its reader in all related ways of AI.

5.8 Result or Conclusion with Future Scope

This chapter brought the needful and wider information about the implementation of strategic business management about humanoid robots including developing robotics and AI companies which will build a scope across the world in the field.

We should always appreciate technology and technologists who are committed to help advance the cutting edge in self-ruling mechanical technology advances. Nevertheless, there are challenges in AI for the future. AI or IOT has given more scope with new vision demanding genuine learning for full-time courses of this new branch to educational institutes. Nevertheless, we should think in the direction "will we become slaves of AI somewhere?" Otherwise AI in the form of robotics is always helpful and useful by doing their untidy strength.

References

1. E.K.K. Rich et al., *Artificial Intelligence* (McGraw Hill Education, New Delhi, 2014), p. 3
2. J. Stuart, P. Norvig, et al., *Artificial Intelligence A Modern Approach* (Pearson, New York, 2017), pp. 17–33
3. J. Bindra, Hindustan, 2020, Haldwani, p. 10
4. Josh, Jobs profile in India, 2020, Dainik Jagaran, Haldwani, p. 17
5. T.H. Davenport, *The AI Advantage: How to Put the Artificial Intelligence Revolution to Work* (The MIT Press, Cambridge, Massachusetts, 2019)
6. The AI Advantage: How to Put the Artificial Intelligence Revolution to Work (Management on the Cutting Edge) Paperback–Illustrated, 2019. Available at: https://www.amazon.com/gp/product/0262538008. Accessed 7 Oct 2020)
7. Artificial Intelligence. What is Artificial Intelligence? How does AI work?, 2020. https://builtin.com/artificial-intelligence. Accessed 7 October 2020
8. Reasoning, 2020. https://www.britannica.com/technology/artificial-intelligence/Reasoning. Accessed 1 October 2020
9. 31 robotics companies on the forefront of innovation, 2020. https://builtin.com/robotics/robotics-companies-roundup. Accessed: 7 Oct 2020
10. Nouvelle artificial intelligence computer science, 2020. https://www.britannica.com/technology/nouvelle-artificial-intelligence. Accessed 1 Oct 2020
11. Top 10-robots-man-made-reasoning, 2017. https://www.designboom.com/innovation/top-10-robots-man-made-reasoning. Accessed 1 Oct 2020
12. 10 of Japan's top robotics companies leading the world into 2020, 2020. https://www.rocketspace.com/corporate-innovation/10-of-japans-top-robotics-companies-leading-the-world-into-2020. 22 Oct 2020
13. Stanford's search and rescue snake robot is a slithering success, 2020. https://www.designboom.com/technology/stanford-snake-robot-07-20-2017/. Accessed 1 Oct 2020
14. Honda unveils E2-DR, its first humanoid disaster-relief robot, 2020. https://eandt.theiet.org/content/articles/2017/10/honda-unveils-e2-dr-its-first-humanoid-disaster-relief-robot/. Accessed 9 Oct 2020
15. G. Garcia, Artificial Intelligence in Japan, 2020
16. Government of Japan. Innovating the future. How Japan uses AI and robotics to solve social issues and achieve economic growth, 2020. https://hbr.org/sponsored/2020/02/how-japan-uses-ai-and-robotics-to-solve-social-issues-and-achieve-economic-growth. Accessed 21 Oct 2020

17. 7 humanoid robots which were made in India, and their success stories, 2018. https://analyticsindiamag.com/7-humanoid-robots-which-were-made-in-india-and-their-success-stories/. Accessed 30 Oct 2020
18. Panache, South Korea changes how people dine: AI robots negate contact with staff, serve customers, 2020. *The Economic Times*, Sep 15, 2020. https://economictimes.indiatimes.com/magazines/panache/south-korea-changes-how-people-dine-ai-robots-negate-contact-with-staff-serve-customers/articleshow/78126725.cms. Accessed 30 Oct 2020
19. Top Artificial Intelligence companies in India, 2020. https://www.goodfirms.co/artificial-intelligence/india. Accessed 9 Oct 2020
20. B. Sukheja, South Korea: Robot with artificial intelligence helps fight COVID-19 spread', 2020. https://www.republicworld.com/world-news/rest-of-the-world-news/robot-with-artificial-intelligence-helps-fight-covid-19-spread.html. Accessed 30 Oct 2020
21. These are the countries with the highest density of robot workers (2020). Available at: https://www.weforum.org/agenda/2020/09/countries-comparison-robot-workers-robotics-change-tech-manufacturing/ (Accessed:30 October 2020)
22. I. Zueco, Will AI solve your workplace safety problems? EHS Blog, 2020. https://www.pro-sapien.com/blog/ai-increase-safety-productivity/. Accessed 8 Oct 2020
23. Neelie, 5 fascinating uses of AI in recruitment, 2018. https://harver.com/blog/uses-ai-in-recruitment/. 8 Oct 2020

Chapter 6
Artificial Intelligence: Technologies, Applications, and Policy Perspectives. Insights from Portugal

Maria José Sousa, Francesca Dal Mas, Gabriel Osório de Barros, and Nuno Tavares

6.1 Introduction

Today, technology profusion and innovation are enormous [1] and fast, able to change the global scenario including businesses and how they operate, how they manufacture and communicate with the consumer, generating new business models [2, 3]. The most critical contemporary technologies incorporate artificial intelligence (AI) [4–7].

Lately, there has been an increasing interest in AI from both academics and practitioners [8, 9]. AI is connected to machines' capacity to imagine and process like real people, which means having the ability to read, interpret, and determine in a logical and intelligent way [10]. Some of the technologies associated with AI are deep learning (DL), machine learning (ML), and natural language processing (NLP).

DL is intended when computers use complex algorithms to emulate the human brain's neural network and practically learn an area of information without supervision [11].

M. J. Sousa
Instituto Universitário de Lisboa, Business Research Unit, Lisbon, Portugal
e-mail: maria.jose.sousa@iscte-iul.pt

F. Dal Mas (✉)
Department of Management, Lincoln International Business School, University of Lincoln, Lincoln, UK
e-mail: fdalmas@lincoln.ac.uk

G. Osório de Barros · N. Tavares
Gabinete de Estratégia e Estudos do Ministério da Economia e Transição Digital, Lisbon, Portugal
e-mail: gabriel.barros@gee.gov.pt; nuno.tavares@gee.gov.pt

P. M. Jeyanthi et al. (eds.), *Decision Intelligence Analytics and the Implementation of Strategic Business Management*, EAI/Springer Innovations in Communication and Computing, https://doi.org/10.1007/978-3-030-82763-2_6

ML includes computers that learn from, often with a minimum of programming, the data and knowledge that are applied to them [12], reaching the results autonomously (for instance, the recommendations on Amazon). ML allows a mathematical model to be built from data, including a large number of unknown variables. With training data sets, the parameters are configured during the learning process. The various methods of ML are divided into three categories: human-supervised learning, unsupervised learning, and reinforcement-unsupervised learning. Such groups include various techniques, among others deep learning and neural networks.

NLP refers to ML methods employed to track down patterns in extensive data sets that recognize the natural language [13]. The study of feelings, where algorithms will search for social networking trends to investigate the consumers' feelings and attitude towards particular products, brands, or goods, is an example.

New advanced analytical tools have the potential to automatically classify market behaviors and predictions, incorporating analytical frameworks into operations affecting key decision-making capabilities, with a relevant effect on the architecture of organizations where new functions arise [14]. AI profoundly affects the way firms make their business strategies, as it impacts both the organization's internal resources and the way the business environment shapes its forces. While AI can be considered part of a company's structural capital [15], it can represent a way for organizations to boost their capabilities and competitive advantage [16]. AI follows automation, cloud computing, and the Internet of Things (IoT), to empower advanced machines, more intelligent factories, more innovative ecosystems, leading to the creation of new business models [17–20].

AI is driven by the convergence of virtually unlimited cloud computing capacity, digitization, and breakthroughs on how machines can use such data to understand and reason as well as humans do. Organizations are more and more willing to adopt modern AI technology such as ML. Among the goals, companies aim to speed production, boost their operating efficiencies, maximize equipment efficiency, eliminate garbage, and keep maintenance fees under control [17].

AI innovations are now opening up a mixed workforce in which humans and computers operate together. The aim is to find the perfect combination of human and machine cooperating to allow workers to benefit from technology's power [21]. According to the global market intelligence firm IDC, by the end of 2020, 60 per cent of plant floor workers at G2000 manufacturers will collaborate with technologies that empower automation, such as additive manufacturing, simulation, AI, robotics, and augmented and virtual reality. Still, there is an open debate about if and how much AI can replace human intelligence. The dialogue involves, in particular, all the sectors in which AI technologies are being adopted fast, disrupting the business processes and the outcomes. While most of the literature highlights how AI cannot replace human intelligence, almost all studies agree that it can support and enhance several strategic processes, including decision-making [22–24].

As said, AI is a technology which has been disrupting and boosting the performance and outcomes of the manufacturing and services sectors, and more is about to happen in the next few years [25]. This study carries an inclusive and general AI analysis, underlining the main theoretical frameworks and experiences in practice, providing a comprehensive review of AI from an industrial perspective. In trying to achieve this aim, the chapter provides first a systematic literature review to define the trends and applications of AI, followed by a reflection about the potential policy implications. The case of Portugal is taken as an example of what is going on in the field, and which challenges come next. This study has been conducted together with the Portuguese Ministry of Economic to analyze the technologies, applications, and policy perspectives of AI, focusing on Portugal.

6.2 Methodology

In trying to map the phenomenon and its main characteristics, a systematic literature review has been conducted starting from a search on the scientific databases Web of Science and Scopus [26]. This methodology was chosen based on Edmondson and McManus [27] studies and ideas about the best methods to analyses the problem of the technologies and application of AI in the industrial and service scenario.

The chapter addresses three main research questions (RQs), namely: RQ1: What are the main AI technologies applied to industry and services?; RQ2: Which the main applications of AI to the industry and services?; and RQ3: Which are the main public policies issues regarding AI in the European Union and Portugal?

The selected keywords were "artificial intelligence" and "technologies" and "applications." The search was limited to the timeframe 2015-(mid) 2020 since the interest towards the topic has been growing in the last 5 years, as well as the level reached by the technology and its applications in the industry. The first search was based only on "artificial intelligence" keyword, and the result was 125,957 articles. It was then necessary to restrict the search key to make it more aligned to the study's goals. Therefore, several exclusion criteria were included: the other keywords considered "technologies" and "applications," a restriction of the time frame to the last 5 years; only articles published in academic reviews; articles written in English, and full access. The final results lead to 314 research articles, which were analyzed.

6.3 AI: Themes, Sectors, and Applications

In addressing RQ1: What are the main AI technologies applied to industry and services?, the analysis of the 314 selected scientific articles allowed identifying the main research areas on AI and also the main sectors where it is applied. The trends of AI are based on a profusion of platforms, applications, and tools. Furthermore, from the analysis of the scientific studies about AI, the main themes and technologies under investigation are ML, predictive analytics, artificial neural network, DL, robotics, and AI. Table 6.1 summarizes the main results.

Table 6.1 Main themes in artificial intelligence studies

Keywords	Number of publications
Artificial Intelligence	134
Artificial neural network	24
Big data	9
Blockchain	2
Data science	3
Deep learning	22
Machine learning	61
Predictive analytics	37
Robotics	19
Smart cities	3
Total	314

Source: The Authors' analysis on Web of Science and Scopus, 2020

Table 6.2 Sectors being studied in artificial intelligence studies

Research areas	Number of publications
Business and management	8
Citizen privacy and law	5
Computers and engineering	79
Emergent technologies	32
Energy	30
Health	100
Industry	13
Public sector	3
Safety and environmental	39
Smart cities	4
Tourism	1
Total	314

Source: The Authors' analysis on Web of Science and Scopus, 2020

AI makes it possible to build algorithms known as artificial neural networks, inspired by the human brain's role. ML and DL's fields are more used are areas of AI technology such as machine vision, autonomous vehicles, automated text generation, face recognition (i.e. the use of mobile or personal computer facial recognition to access devices), and so on. AI makes software to be used in machine vision, voice recognition, and NLP cases even more straightforward.

AI innovations have many consequences for the internet business, computing and engineering, and many other areas, such as manufacturing, healthcare (which appears the most investigated sector), agriculture and automotive. Table 6.2 summarizes the main results.

More in detail, AI and ML are already showing a massive impact on healthcare and surgery [28]. The AI-based applications can support clinical and surgical decision-making, understand and interpret medical data, formulating a diagnosis without direct human intervention [29]. These applications are applied not only in the diagnosis phases but also in the treatment definition, image-guided surgery, drug development, personalized medicine, and patient monitoring and care [30]. Several clinical disciplines are involved in the AI revolution. Still, disruptive impacts are expected in radiation diagnosis [31, 32], eye-testing apparatus [33], ultrasonic diagnosis [34], diagnostics [35], computer-aided surgery [29], surgery [36], X-ray, and radiation therapy [37].

AI and ML also significantly impact the banking industry [38, 39]. The use of these new technologies simplifies banking operations. As an example, many banks and financial institutions have started to employ intelligent virtual assistants to enhance post-sales and customer support. Moreover, AI systems are used to prevent fraud and check possible dangers to clients while shopping in online stores.

Concerning the e-commerce industry as a retail market, the use of AI has included chatbots, helpers designed on AI technologies, smart logistics, and algorithms to monitor, assess, forecast, and interpret the actions of consumers [40, 41].

Many businesses often use operational automation to reach higher productivity standards and reduce transportation costs [42]. ML assists businesses in market forecasts, product search rating, product and deal suggestions, detect fraud, translations, and many more practices [43].

Regarding our RQ2: Which the main applications of AI to the industry and services?, predictive analytics stands as the number one AI use. It employs statistical methods and calculation to identify if and when events and outcomes are likely to happen and which counteractions can be taken. Practical examples include the ability to forecast if and when a wagon or pick-up is expected to crash or tear down and when it may be stuck in traffic or delayed because of unfavorable weather conditions [42]. The second most popular application involves real-time operations management, used to improve internal efficiencies within the company [44]. The possibility to enhance and monitor customer services [45] and customer insight and experience comes next, with dedicated new marketing operations by analyzing the outcomes of purchases. Companies are then able to map and understand the new habits, trends, behaviors of the customers [46]. Additional uses include risks management [47], research and development [25], supply chain, logistics, and procurement [48], human resource management [49], fraud prevention and detection through cybersecurity [50], knowledge creation, pricing, and social engagement [51].

AI also affects education [9]. The literature focuses on the contributions of AI mainly to improve learning opportunities for students and the management of the learning process [52–55]. AI solutions prove to contribute to equitable and inclusive access to education, providing opportunities, e.g., for people with disabilities and those living in isolated communities to access appropriate learning paths, through holograms and robotics. AI's contribution to education can support students who are not physically in the same location, allowing teachers and mentors to monitor asynchronous discussion groups, boosting engagement and outcomes. A secondary output allows better management of the flows, including grading.

All in all, AI improves strategic decision-making in various fields. Business decision-making requires analytical and intuitive thinking. AI solutions proved to integrate, but not to substitute, human explicit and tacit knowledge, supporting strategic decision-making [56]. Practical examples include healthcare and surgery [57, 58]. According to the literature, surgical decision-making is affected by time constraints (especially in trauma and emergency surgery), bias, uncertainty, complexity, decision fatigue, and hypothetical-deductive reasoning, leading to disservice and potential mistakes. In the hypothetical-deductive framework that characterizes surgical decision-making, the surgeon or surgical team should assess the initial patient condition by developing a possible diagnosis list. Such a list should then lead to a series of tests or therapies. In the traditional surgical scenario, it is the surgeon's ability or guesswork to create a comprehensive list of all likely or unlikely diagnoses and life-threatening factors [59], also taking into consideration the strengths and weaknesses of available trials or therapies. Once the best-fitting diagnosis is identified, the surgeon must use good judgment to formulate a surgical or treatment plan. All the phases along the road pose variability and possibilities for errors or biases [60]. Traditional clinical decision aid systems and tools are often compromised by time-consuming manual data entry and suboptimal accuracy. AI has the power to address these weaknesses. AI-based tools and technology that are correctly implemented can facilitate surgical decision-making by endorsing the decision to run the procedure, the informed consent process, the recognition and mitigation of eventual risk factors, the mapping and care of potential side effects, and shared decisions on the available resource [57, 61]. Still, AI solutions and technologies can support and improve, rather than replace, human decision-making [62] recalling the need to consider and train AI as a means for augmentation (boosting human's capabilities) rather than automation (replacing them).

6.4 Policy Reflections About Fostering Artificial Intelligence

The findings reveal that research on AI has advanced rapidly over the past years, and it is increasing again. In addressing our RQ3: Which are the main public policies issues regarding AI, with a major focus on the European Union and Portugal, we should highlight how while the technological applications and their outcomes on a variety of industries and sectors are undoubtedly impressive and measurable, concerns from the society have arisen [63, 64]. Simultaneously, two opposing views about these impacts have emerged. A more optimistic view focuses on the fact that AI could dramatically enhance productivity and outcomes in various fields, impacting people's lives positively. Applications like those on healthcare could lead to better patients' outcomes, like more precise diagnosis and safer surgeries and treatments. Fraud prevention could save people's money, and personalized customers' services may enhance the users' experience. On the contrary, a more pessimistic view stresses that the fast developments and applications in AI could impact society for the worse, reducing jobs and invading people's privacy, raising thus ethical concerns and legal issues.

Policy choices are likely to play a critical role, regardless of whether one takes a more negative or positive view on the topic. Second, regulatory policies will definitely impact the pace of technological diffusion and the shape in which technology takes. Third, some measures serve as protections for AI's future detrimental impacts on labor markets and competition issues [65].

As for policies that will impact the propagation of AI, one might point to ML's example. This technology is vulnerable to using vast quantities of data to make predictions feasible. In specific environments, the main limitation of AI is the right to collect valuable data, much of which is subject to privacy issues. In this way, the privacy policy has a significant effect on organizations' abilities to develop and enforce AI [66]. Very little security of privacy ensures that customers will not be inclined to engage in business purchases where they are exposed to their data and information. Too much regulation of privacy provides that businesses cannot use data to innovate. Although the latest analytical study does not directly concentrate on AI, the evidence to date shows that most government-mandated legislation on privacy is likely to slow down the adoption and advancement of technology, indicating a trade-off between the right to privacy and the pace of innovation [67]. This implies that any AI-focused government policy should weigh data producers and consumers' potentially competing interests, particularly concerning privacy, primarily to support a local AI industry. Perhaps more than any other legislation, privacy laws are likely to impact the pace and course of AI implementation in operation.

Liability and accountability rules may also affect the implementation of AI [68]. Companies would be less likely to invest in producing AI products without strict and transparent liability laws. The recent development of autonomous cars offers a helpful example. In the production of a self-driving vehicle, a variety of different firms would be interested. Without explicit cut rules on responsibility, however, anyone can hesitate to invest. Suppose autonomous cars can save lives, allowing a more secure driving. Should non-autonomous vehicle manufacturers be kept to higher standards than what is mandated under current law? The dissemination of safer technology will intensify this. In contrast, if the rise in accountability was mainly on emerging technologies, then diffusion would slow down. As for all other technology, more scientific funding, well-balanced intellectual property legislation, and the freedom to innovate in a protected manner would make development quicker.

The same liability issue concerns the evolution and adoption of AI-based surgical robots. Such robots can already independently carry out parts of surgical operations like, for instance, performing intestinal anastomoses more precisely and faster than experienced surgeons [69]. The benefits of such robots have already been stated and measured, such as a more precise visualization of the operating field, better movements thanks to the articulating tools, the elimination of vibrations and fewer medical errors [29], leading to better outcomes both for the surgical patients as well as the hospital or institution adopting such a technology [70]. However, today, the entire decision-making process is the full responsibility of the surgeon in charge [71]. This is the main reason limiting the spread of AI-based surgical robots [72], and this issue will remain until further regulations.

As for policies that address AI outcomes, one could point to the most common concerns, the impact on jobs [62]. Although an argument can be made about the fact that technology is often put into place to substitute human activity in tedious and dangerous undertakings, increasing safety and productivity [45, 62], some have raised concerns about the need to find alternate sources of meaning. Moreover, the widespread use of AI may further strengthen current wealth distribution trends. In other words, the development of AI may come at the cost of further inequality. It is likely to be skill-biased if AI is like other forms of information technology. Skilled people who are still doing reasonably well would be those who profit most from AI [73]. Such persons are also more likely to own computers and technology. The social safety net is linked to measures to counter the effects of AI for inequalities. The AI background is not unusual in weighing the risks and advantages of social services, from egalitarian taxes to healthcare coverage. However, others have floated relatively ambitious proposals to deal with the possible rise in inequalities, such as a levy on machines and a universal basic income.

In the shorter term, the change could mean temporary redundancy for many employees if AI diffuses broadly. Some authors emphasize a short-to-medium-term disparity between skills and technologies. This suggests that policy planning should consider both economic cycles and education policy in advance of the dissemination of AI. Technology-driven dismissals based on place and time are not unique to AI. They became a feature of the modernization of factories and farm mechanization. There are also unanswered questions concerning school reform. If computers can execute technology-related prediction functions, can education programs concentrate on cognitive skills and the humanities? Should school programs evolve to rely more on adults? How do the skills available as AI spreads vary from the skills currently offered by the education system? It is essential to answer all these unanswered questions through effective policies and services.

Another policy question relates to whether AI will lead to monopolization in some industries. The leading AI companies are prominent in revenue, profits, and market capitalization [74]. This has contributed to a rise in antitrust scrutiny by governments of leading technology companies (especially the European Commission). These organizations' position as forums, not their use of AI per se, is the object of much of this antitrust scrutiny. The meaning of data is the function that makes AI different. Companies can develop better AI with more info. If this leads to economies of scale and the opportunity for monopolization depends on whether a slight lead provides a positive feedback loop and a long-run benefit early in the growth cycle.

While AI is like other technology in many ways, it looks unique in a few crucial dimensions. In specific, AI is both a general-purpose application and an invention' approach. The consequences concern returns on expenditures like AI regulation. The cost of suboptimal policy design is likely to be considerably higher than for other innovations due to the scope of implementations or the advantages of optimal policy are more important.

6.5 EU Artificial Intelligence Strategy 2030

The EU Artificial Intelligence Strategy 2030 has been validated, and it is running. It foresees that AI will be included in educational materials for pupils and grown-ups, either embedded into the public administration services or as SMEs technologies. The aim is to investigate AI's various economic and societal potentials, as well as its application in areas such as renewable energy networks, cities, metropolis, and communities, forests, and oceans, connectivity, self-driving vehicles, and healthcare services. Its growing popularity will help to advance scientific research in the future.

This Strategy's plan clarifies that AI's investment brings more opportunities and jobs, guaranteeing all workers' inclusion impacted by this new reality. While creating a new world of possibilities, AI will not replace human beings. Promoting a better environment, encouraging AI talents and digital brains, generating new jobs and building an AI service economy stand as the main pillars of the AI strategy. The aim is to secure AI niche markets through key specialized services, leading to further innovations through AI science, delivering better public services to people and businesses.

In this context, Portugal has been selected as a living lab for the experimentation of new AI developments, implementing the National Strategy for Artificial Intelligence, based on the InCoDe2030 program, under the Government of Portugal strategy and policies and operationalized by these organizations: Portuguese Science and Technology Foundation (FCT), Ciência Viva, the Portuguese Innovation Agency (ANI), the Portuguese Agency for Administrative Modernisation (AMA). The Action Plan includes seven action lines, concerning inclusion and education by disseminating generalist knowledge, qualification, and specialization, the definition of thematic areas for research and innovation in European and international networks, the development of the public administration and its modernization, the identification of specific areas of specialization with global impact, the development and support of regions in European and international networks, and addressing the challenges in terms of ethics and safety.

6.6 Portugal Artificial Intelligence Overview

Notwithstanding the increasing significance of AI, there are substantial holes in the Portuguese case's quantitative knowledge. Most of the current experience, in particular, refers to studies focused on the collection of data from different stakeholders in the form of surveys. CIONET Portugal researched the effect of AI on Portuguese firms belonging to various industries. One of the most critical pieces of evidence from the research shows that 34.6% of participants would use an AI approach in less than 12 months. As of now, 39% of participants state their company is employing AI in everyday activities. Following this report, participants suggest that ML and chatbots will be the AI technologies that will gain more success in the industry and service sectors. ML stands as the most accepted AI tool. 94% of organizations say they will likely implement this approach. Still, participants claim as AI may impact their organizations by, for example, supporting the automation of facilities, IoT, or even hospital diagnostic tools.

The uncertainty that employment will shrink increases with the growth of AI. Thus, a positive impact on Portuguese employability can be expected, with potential growth of 15.1%, despite the number of jobs lost because of AI applications.

A different piece of research on a sample of Portuguese firms (namely, Artificial Intelligence in Europe: How Leading Companies Benefit from AI - Perspectives 2019 and Beyond) highlights how ML and Smart Robotics technologies can represent significant opportunities for Portuguese companies.

Portuguese companies appear to be late adopters of AI compared to other European countries and, considering a ratio of AI maturity, Portuguese organizations stand behind the European average. While 82% of Portuguese firms declare they would soon try or employ some AI-based applications, the remaining 18% are not yet thinking to introduce AI into their plants. Therefore, results show that the maturity level is still far away, and policies and actions are needed to reach it.

In certain companies/functions, the diffusion of AI among Portuguese entities is depleted. However, all company activities are protected by AI software. Nevertheless, the investigation resulted in some surprising findings, as 36% of firms do not employ AI-based applications in the management or operations. In line with European trends, the involvement of Portuguese companies which apply AI in their fields appears in this way: Technology and Digital (45%), Client Support (36%) and Research, Development, and Product Creation (32%).

The problem with sector delimitation is one of the fundamental issues concerning the AI phenomena' mediation component. In this sense, the Techno-Economic Section (TES) methodology was used in a recent piece of research carried on by the European Union. The methodology's first move was to describe the TES (AI, in this case) boundaries, defining players who rely on AI as a primary or secondary operation. Such relevant actors are classified as R&D centers, academic entities, and firms engaged in one or more of the following activities: research processes, manufacturing and marketing industries, and relevant services relating to AI. In the EU background, the United Kingdom, Germany, Spain, Italy, France, and the Netherlands have the most significant amount of organizations. In exchange, the emphasis is on nations such as Cyprus, Bulgaria, Estonia, and Malta, considering the investments as a percentage of their recorded GDP. Portugal holds an intermediate position when considering the quantity of parties and its portrayal compared to the GDP.

On the other side, the same European Commission report mapped the foremost players by country in the 2009–2018 period, who engaged in FP7 and H2020 AI-related research projects. We may observe an equitable distribution among the biggest and highly successful researching Nations. In Portugal, only two of the listed AI organizations got a sponsorship successfully.

6.7 Portuguese AI Case Studies

In recent years, there has been a growing number of research projects and several AI start-ups in Portugal. These include companies such as Feedzai, James, Heptasense, Jungle.ai and Unabel. Among these, Feedzai is undoubtedly one of the most recognized Portuguese start-ups. This firm, which was born in Coimbra, put AI to work on prevention and fraud. Currently, the Feedzai team is working with some of the largest banks in the world. Feedzai has successfully developed an intelligent platform that absorbs and transforms various data streams and fraud information into any channel. The platform enriches data to create hyper-granular risk profiles, while ML processes events and transactions in milliseconds. They then provide explainable AI by adding a human-readable semantic layer to the underlying machine logic.

6.8 Artificial Intelligence Policy Ramifications

AI research has progressed steadily over the past years. Two perspectives can be found in the literature. A more cynical view emphasizes how rapid developments in AI can change society with substantial job losses, arising complex ethical concerns related to the implementation and usage of these technologies. In contrast, a more optimistic perspective focuses on the fact that AI can lead to a better quality of life.

Policy decisions on the implementation of AI in industry and society are likely to play an important role, namely: (a) regulatory policies that can lead to increasing the pace of technology dissemination; (b) privacy policies have a direct effect on organizations' ability to develop and introduce AI. However, a high degree of privacy regulation suggests that companies would have trouble using data to innovate and potentially slow the acceptance and innovation of technology.

This study offers information to policy-makers to help them decide to establish an ethical structure for a more trustworthy, ethical, open, and impartial use of AI. There will also be a strengthening of trust with efforts to consider new standards and norms. It will also be essential to establish policies on ownership of intellectual property rights in outputs generated by AI, liability, and data privacy. Future liability decisions will also impact the distribution of AI in this regard [68], as it affects organizations' faith in investing in AI technologies production.

The decision to develop inequality policies is also nuclear, as AI is likely to be ability biased. People need to adapt to the digital economy's challenges, and policy-makers also need to consider AI in both business cycles and education policy. Policy choices should help corporations explore innovations and upskilling/reskilling the workforce. The digital and data-driven environment pushes forward all aspects of the economy, and policy decisions will speed up adoption, making AI more available and affordable.

6.9 Conclusion

Because of new advances in AI, the effects of sudden and large-scale automation have been the focus of interest lately. In tandem with computers' growing computing power, modern algorithms have made it possible for robots to take care of tasks that only humans can perform until now.

While there are many opportunities in terms of applications covering several different areas, many issues exist, notably data protection, work losses, and deficiencies in the skills required. Looking ahead, the capacity to implement more advanced technology will be missing in less efficient businesses. In turn, from a competitive advantage point of view, the more efficient firms can result, the more value they can get, particularly in a digital world where intangible firm-specific assets look crucial. This feature seeks to disrupt the dynamism and competition in the industry and may potentially impact productivity. In this regard, mounting signs of rising markups and market concentration and decreasing business entries and exits, notably in digitally intensive industries, are alarming, highlighting the possible effect of AI-related technology on organizations' strategic management and strategy-making processes.

Big data on job qualifications and job conditions can be elaborated and analyzed by AI. To direct individuals and employees to make well-informed job decisions, local labor market knowledge and skills, anticipation programs would be necessary. Jobs and training programs, both at the national and local level, will adapt to the next frontiers on employment by identifying demands more efficiently and educating people with valuable expertise. Thus, strategic choices regarding AI involve organizations or industries and countries and policy-makers, who are responsible for creating ad-hoc policies to shape the future of the economy, the business environment, and society to benefit from new technologies.

In all, while AI-based technologies have undoubtedly the power to bring benefits for both companies and society, a need for dedicated policies emerges, both from a national and an international level. While practitioners concentrate on the use and possible applications of such a disruptive technology, scholars should investigate the major outcomes and concerns. Academic studies and professional reports should guide decision-makers in ruling such a challenging phenomenon. Several paradoxes emerge, like the request for privacy in contrast with data collection to foster innovation.

Like all pieces of research, our study has limitations. The continuous and rich production of scientific documents and practitioners' sources on an emerging and actual topic like AI may limit the validity of our findings and considerations in the future. Still, while our research is mainly based on scientific sources, a professional perspective should be added, enriching the phenomenon's view and the considerations on policy implications. Comparing more countries to Portugal may then add valuable insights to the whole topic.

References

1. C. Bagnoli, F. Dal Mas, M. Massaro, The 4th industrial revolution: Business models and evidence from the field. Int J E-Services Mob Appl. **11**(3), 34–47 (2019)
2. C. Nielsen, M. Lund, M. Montemari, F. Paolone, M. Massaro, J. Dumay, *Business Models: A Research Overview* (Routledge, New York, 2018)
3. C. Bagnoli, M. Massaro, F. Dal Mas, M. Demartini, Defining the concept of business model: Searching for a business model framework. International Journal of Knowledge and System Science **9**(3), 48–64 (2018). https://doi.org/10.4018/IJKSS.2018070104
4. M.J. Sousa, R. Cruz, Á. Rocha, M. Sousa, Innovation trends for smart factories: A literature review, in *New Knowledge in Information Systems and Technologies WorldCIST'19 2019 Advances in Intelligent Systems and Computing*, ed. by Á. Rocha, H. Adeli, L. Reis, S. Costanzo, (Springer, Cham, 2019), pp. 689–698
5. M.J. Sousa, A. Pesqueira, C. Lemos, M. Sousa, A. Rocha, Decision-making based on big data analytics for people Management in Healthcare Organizations. J. Med. Syst. **43**(9), 290 (2019)
6. M.J. Sousa, A. de Bem Machado, Blockchain technology reshaping education contributions for policy, in *Blockchain Technology Applications in Education*, ed. by R. C. Sharma, H. Yildirim, G. Kurubacak, (IGI Global, Hershey, PA, 2020), pp. 113–125
7. A. Pesqueira, M. Sousa, A. Rocha, Big data skills sustainable development in healthcare and pharmaceuticals. J. Med. Syst. **44**, 197 (2020)
8. K. Toniolo, E. Masiero, M. Massaro, C. Bagnoli, Sustainable business models and artificial intelligence. Opportunities and challenges, in *Knowledge, People, and Digital Transformation: Approaches for a Sustainable Future*, ed. by F. Matos, V. Vairinhos, I. Salavisa, L. Edvinsson, M. Massaro, (Springer, Cham, 2019), pp. 103–117
9. M.J. Sousa, F. Dal Mas, A. Pesqueira, C. Lemos, J.M. Verde, L. Cobianchi, The potential of AI in health higher education to increase the students' learning outcomes. TEM Journal **10**(2), 488–497 (2021). https://doi.org/10.18421/TEM102-02
10. P. Wang, On defining Artificial Intelligence. J. Artif. Gen. Intell. **10**(2), 1–37 (2019)
11. J. Schmidhuber, Deep learning in neural networks: An overview. Neural Netw. **61**, 85–117 (2015)
12. W.J. Murdoch, C. Singh, K. Kumbier, R. Abbasi-Asl, B. Yu, Definitions, methods, and applications in interpretable machine learning. Proc. Natl. Acad. Sci. U. S. A. **116**(44), 22071–22080 (2019)
13. S. Bird, E. Klein, E. Loper, *Natural Language Processing with Python: Analyzing Text with the Natural Language Toolkit* (Sebastopol, O'Reilly Media, Inc, 2009)
14. D. Adebanjo, P.L. Teh, P.K. Ahmed, The impact of supply chain relationships and integration on innovative capabilities and manufacturing performance: The perspective of rapidly developing countries. Int J Prod Res **56**(4), 170 (2018). https://doi.org/10.1080/00207543.2017.1366083
15. H. Harlow, Developing a knowledge management strategy for data analytics and intellectual capital. Meditari. Account. Res. **26**(3), 400–419 (2018). https://doi.org/10.1108/MEDAR-09-2017-0217
16. S.L. Wamba-Taguimdje, S. Fosso Wamba, R. Kala Kamdjoug Jean, C.E. Tchatchouang Wanko, Influence of artificial intelligence (AI) on firm performance: the business value of AI-based transformation projects. Bus. Process. Manage. J. **26**(7), 189 (2020). https://doi.org/10.1108/BPMJ-10-2019-0411

17. C. Bagnoli, A. Bravin, M. Massaro, A. Vignotto, *Business Model 4.0* (Venezia, Edizioni Ca' Foscari, 2018)
18. F. Dal Mas, M. Massaro, J.M. Verde, L. Cobianchi, Can the blockchain lead to new sustainable business models? J Bus Model. **8**(2), 31–38 (2020)
19. F. Dal Mas, G. Dicuonzo, M. Massaro, V. Dell'Atti, Smart contracts to enable sustainable business models. A case study. Manag. Decis. **58**(8), 1601–1619 (2020)
20. M. Del Giudice, Discovering the Internet of things (IoT) within the business process management: a literature review on technological revitalization. Bus. Process. Manag. J. **22**(2), 263–270 (2016)
21. M. Massaro, S. Secinaro, F. Dal Mas, V. Brescia, D. Calandra, Industry 4. 0 and circular economy: An exploratory analysis of academic and practitioners' perspectives. Bus. Strateg. Environ. **30**(2), 1213–1231 (2021)
22. G. Briganti, O. Le Moine, Artificial intelligence in medicine: Today and tomorrow. Front. Med. **7**, 1–6 (2020)
23. M.H. Stanfill, D.T. Marc, Health information management: Implications of Artificial Intelligence on healthcare data and information management. Yearb. Med. Inform. **28**(1), 56–64 (2019)
24. C. Paton, S. Kobayashi, An Open Science approach to Artificial Intelligence in healthcare. Yearb. Med. Inform. **28**(1), 47–51 (2019)
25. B.H. Li, B.C. Hou, W.T. Yu, X.B. Lu, C.W. Yang, Applications of artificial intelligence in intelligent manufacturing: a review. Front Inf Technol Electron Eng. **18**(1), 86–96 (2017)
26. M. Massaro, J.C. Dumay, J. Guthrie, On the shoulders of giants: Undertaking a structured literature review in accounting. Accounting, Audit Account J. **29**(5), 767–901 (2016)
27. A.C. Edmondson, S.E. McManus, Methodological fit in management field research. Acad. Manage. Rev. **32**(4), 1155–1179 (2007)
28. A.L. Beam, I.S. Kohane, Big data and machine learning in health care. JAMA **319**(13), 1317–1318 (2018)
29. F. Dal Mas, D. Piccolo, L. Cobianchi, L. Edvinsson, G. Presch, M. Massaro et al., The effects of artificial intelligence, robotics, and industry 4.0 technologies. insights from the healthcare sector, in: Proceedings of the first European Conference on the impact of Artificial Intelligence and Robotics. Academic Conferences and Publishing International Limited, 2019, pp. 88–95
30. A. Becker, Artificial intelligence in medicine: What is it doing for us today? Heal. Pol. Technol. **8**(2), 198–205 (2019)
31. M.P. McBee, O.A. Awan, A.T. Colucci, C.W. Ghobadi, N. Kadom, A.P. Kansagra, et al., Deep learning in radiology. Acad. Radiol. **25**(11), 1472–1480 (2018). https://doi.org/10.1016/j.acra.2018.02.018
32. R.C. Mayo, J. Leung, Artificial intelligence and deep learning – Radiology's next frontier? Clin. Imaging **49**, 87–88 (2018)
33. R. Kapoor, S.P. Walters, L.A. Al-Aswad, The current state of artificial intelligence in ophthalmology. Surv. Ophthalmol. **64**(2), 233–240 (2019)
34. A.F. Gosling, R. Thalappillil, J. Ortoleva, P. Datta, F.C. Cobey, Automated spectral Doppler profile tracing. J. Cardiothorac. Vasc. Anesth. **34**(1), 72–76 (2020)
35. A.M. Bur, A. Holcomb, S. Goodwin, J. Woodroof, O. Karadaghy, Y. Shnayder, et al., Machine learning to predict occult nodal metastasis in early oral squamous cell carcinoma. Oral Oncol. **92**, 20–25 (2019)
36. A. Maubert, L. Birtwisle, J.L. Bernard, E. Benizri, J.M. Bereder, Can machine learning predict resecability of a peritoneal carcinomatosis? Surg. Oncol. **29**, 120–125 (2019)
37. R.F. Thompson, G. Valdes, C.D. Fuller, C.M. Carpenter, O. Morin, S. Aneja, et al., Artificial intelligence in radiation oncology: A specialty-wide disruptive transformation? Radiother. Oncol. **129**(3), 421–426 (2018). https://doi.org/10.1016/j.radonc.2018.05.030
38. M. Jakšič, M. Marinč, Relationship banking and information technology: The role of artificial intelligence and FinTech. Risk Manage. **21**, 1–18 (2019)

39. M.E. Payne, J.W. Peltier, V.A. Barger, Mobile banking and AI-enabled mobile banking: The differential effects of technological and non-technological factors on digital natives' perceptions and behavior. J. Res. Interact. Mark. **12**(3), 328–346 (2018). https://doi.org/10.1108/JRIM-07-2018-0087
40. R.S. Sexton, R.A. Johnson, M.A. Hignite, Predicting internet/ e-commerce use. Internet Res. **12**(5), 402–410 (2002). https://doi.org/10.1108/10662240210447155
41. X. Song, S. Yang, Z. Huang, T. Huang, The application of artificial intelligence in electronic commerce. J. Phys. Conf. Ser. **1302**, 3 (2019)
42. R. Fildes, K. Nikolopoulos, S.F. Crone, A.A. Syntetos, Forecasting and operational research: a review. J. Oper. Res. Soc. **59**(9), 1150–1172 (2008). https://doi.org/10.1057/palgrave.jors.2602597
43. D. Bogataj, M. Bogataj, D. Hudoklin, Mitigating risks of perishable products in the cyber-physical systems based on the extended MRP model. Int. J. Prod. Econ. **193**, 51–62 (2017)
44. J. Wan, J. Yang, Z. Wang, Q. Hua, Artificial Intelligence for cloud-assisted smart factory. IEEE Access. **6**, 55419–55430 (2018)
45. M.H. Huang, R. Rust, Artificial Intelligence in service. J. Serv. Res. **21**(2), 155–172 (2018)
46. N. Syam, A. Sharma, Waiting for a sales renaissance in the fourth industrial revolution: Machine learning and artificial intelligence in sales research and practice. Ind. Mark Manage. **69**, 135–146 (2018)
47. D.D. Wu, S.-H. Chen, D.L. Olson, Business intelligence in risk management: Some recent progresses. Inform. Sci. **256**, 1–7 (2014)
48. H. Min, Artificial intelligence in supply chain management: Theory and applications. Int. J. Logist. Res. Appl. **13**(1), 13–39 (2010)
49. P. Tambe, P. Cappelli, V. Yakubovich, Artificial intelligence in human resources management: Challenges and a path forward. Calif. Manage. Rev. **61**(4), 15–42 (2019)
50. N.F. Ryman-Tubb, P. Krause, W. Garn, How Artificial Intelligence and machine learning research impacts payment card fraud detection: A survey and industry benchmark. Eng. Appl. Artif. Intel. **76**, 130–157 (2018)
51. D. Grewal, A.L. Roggeveen, J. Nordfält, The future of retailing. J. Retail. **93**(1), 1–6 (2017)
52. M. Laanpere, K. Pata, P. Normak, H. Põldoja, Pedagogy-driven design of digital learning ecosystems. Comput. Sci. Inf. Syst. **11**(1), 419–442 (2014)
53. R. Luckin, *Machine Learning and Human Intelligence: The Future of Education for the 21st Century* (UCL IOE, London, 2018)
54. V. Mayer-Schönberger, K. Cukier, *Learning with Big data: The Future of Education* (Boston/New York, Eamon Dolan Book, 2014)
55. M. Montebello, *AI Injected e-Learning: The Future of Online Education* (Springer, Berlin, 2017)
56. L. Metcalf, D.A. Askay, L.B. Rosenberg, Keeping humans in the loop: Pooling knowledge through artificial swarm intelligence to improve business decision making. Calif. Manage. Rev. **61**(4), 84–109 (2019)
57. T.J. Loftus, P.J. Tighe, A.C. Filiberto, P.A. Efron, S.C. Brakenridge, A.M. Mohr, et al., Artificial Intelligence and surgical decision-making. JAMA Surg. **155**(2), 148–158 (2020)
58. P. Mascagni, A. Vardazaryan, D. Alapatt, T. Urade, T. Emre, C. Fiorillo, P . Pessaux, D. Mutter, J. Marescaux, G. Costamagna, B. Dallemagne, N. Padoy, Artificial intelligence for surgical safety: Automatic assessment of the critical view of safety in laparoscopic cholecystectomy using deep learning. Ann. Surg. (2020). https://doi.org/10.1097/SLA.0000000000004351
59. F. Dal Mas, D. Piccolo, L. Edvinsson, M. Skrap, S. D'Auria, Strategy innovation, intellectual capital management and the future of healthcare. The case of Kiron by Nucleode, in *Knowledge, People, and Digital Transformation: Approaches for a Sustainable Future*, ed. by F. Matos, V. Vairinhos, I. Salavisa, L. Edvinsson, M. Massaro, (Springer, Cham, 2020), pp. 119–131
60. F. Dal Mas, D. Piccolo, D. Ruzza, Overcoming cognitive bias through intellectual capital management. The case of pediatric medicine, in *Intellectual Capital in the Digital Economy*, ed. by P. Ordonez de Pablos, L. Edvinsson, (Routledge, London, 2020), pp. 123–133

61. L. Cobianchi, F. Dal Mas, A. Peloso, L. Pugliese, M. Massaro, C. Bagnoli, et al., Planning the full recovery phase: An Antifragile perspective on surgery after COVID-19. Ann. Surg. **272**(6), e296–e299 (2020)
62. M.H. Jarrahi, Artificial intelligence and the future of work: Human-AI symbiosis in organizational decision making. Bus. Horiz. **61**(4), 577–586 (2018)
63. M. Hengstler, E. Enkel, S. Duelli, Applied artificial intelligence and trust—The case of autonomous vehicles and medical assistance devices. Technol Forecast Soc Change **105**, 105–120 (2016)
64. G.A. Montes, B. Goertzel, Distributed, decentralized, and democratized artificial intelligence. Technol Forecast Soc Change **141**, 354–358 (2019)
65. N. Petit, Antitrust and artificial intelligence: A research agenda. J Eur Compet Law Pract. **8**(6), 361–362 (2017)
66. A. Agrawal, J. Gans, A. Goldfarb, Economic policy for artificial intelligence. Innov. Policy Econ. **19**, 139–159 (2018). https://doi.org/10.1086/699935
67. A. Goldfarb, C. Tucker, Shifts in privacy concerns. Am. Econ. Rev. **102**(3), 349–353 (2012)
68. A. Galasso, H. Luo, When does product liability risk chill innovation? Evidence from medical implants. Harvard Business School Working Paper, 2018
69. A. Shademan, R.S. Decker, J.D. Opfermann, S. Leonard, A. Krieger, P.C. Kim, Supervised autonomous robotic soft tissue surgery. Sci. Transl. Med. **8**(337), 337–364 (2016)
70. E. Köse, N.N. Öztürk, S.R. Karahan, Artificial Intelligence in surgery. Eur Arch Med Res. **34**(Suppl. 1), S4–S6 (2018)
71. G. Aruni, G. Amit, P. Dasgupta, New surgical robots on the horizon and the potential role of artificial intelligence. Investig. Clin. Urol. **59**(4), 221–222 (2018)
72. A. Atabekov, O. Yastrebov, Legal status of artificial intelligence across countries: legislation on the move. Eur. Res. Stud. J. **21**(4), 773–782 (2018)
73. H.J. Wilson, P.R. Daugherty, N. Morini-Bianzino, The jobs that artificial intelligence will create. MIT Sloan Manag. Rev. **58**(4), 14–16 (2017)
74. Y.K. Dwivedi, L. Hughes, E. Ismagilova, G. Aarts, C. Coombs, T. Crick, et al., Artificial Intelligence (AI): Multidisciplinary perspectives on emerging challenges, opportunities, and agenda for research, practice and policy. Int. J. Inf. Manage. **2019**, 101994 (2019)

Chapter 7
The Rise of Decision Intelligence: AI That Optimizes Decision-Making

P. Mary Jeyanthi, Dieu Hack Polay, and Tanupriya Choudhury

7.1 The Rise of Decision Intelligence (DI)

- An opinion to decision-making
- Decision intelligence—the pathway
- Framework of DI
- Artificial intelligence (AI) can do and cannot do
- How the AI revolution will transform the business decisions
- Decision intelligence: alignment and optimized the AI decision-making

7.2 The Rise of Decision Intelligence (DI)

Decision intelligence is the discipline of transforming information into intelligence for a better course of action in any kind of environment, technology, and process. Currently, organizations are worrying to get better insight and analysis to utilize the right yield out of artificial intelligence (AI) and machine learning (ML). This is very hectic to rearrange the complex choices of opinion around how the actual

P. M. Jeyanthi (✉)
Bambala Institutional Area Pratap Nagar, Sanganer, Jaipuria Institute of Management Jaipur, Jaipur, Rajasthan, India
e-mail: mary.jeyanthi@jaipuria.ac.in

D. H. Polay
Lincoln International Business School, University of Lincoln, Lincoln, UK
e-mail: DHackpolay@lincoln.ac.uk

T. Choudhury
School of CSE, University of Petroleum and Energy Studies, Dehradun, Uttarakhand, India
e-mail: tanupriya1986@gmail.com

P. M. Jeyanthi et al. (eds.), *Decision Intelligence Analytics and the Implementation of Strategic Business Management*, EAI/Springer Innovations in Communication and Computing, https://doi.org/10.1007/978-3-030-82763-2_7

resources have to be prearranged to plan the responsibilities as per the severity of the workload, and how to manage dynamic limitations to sustain in the market.

In the current scenario, the market features are unpredictable to finalize the decision models. Subsequently, the weakness to catch the vulnerability factors connected to these models' behavior in a business environment. Machine learning techniques play a major role in decision-making processes, the other name can be coined as "decision intelligence" which is rising to make robust decision models in a wide scope of developments.

Decision intelligence is an emergent field that contains a wide range of decision-making strategies to track decision models and processes. The implementation of the decision model provides a construction for organizational decision-making processes with the incorporation of machine learning and optimization algorithms. The valuation of decision intelligence is that decisions depend on the different categorization of patterns to lead the results.

7.3 An Opinion to Decision-Making

"How do you make more effective decisions in the current era? To make effective decisions, do you start with the facts?"

While making decisions, the above questions will often raise in our thought process. To make effective decisions, we have to initially start with opinions. We will try to gather the facts based on the relevant information of data. Then we usually test with our collected opinion with reality.

The first image shows that some square-shaped boxes; once we will concentrate Fig. 7.1b, we will finalize that Fig. 7.1a is the top of the pillar. Once we have start collecting more facts, we will get the relevant opinions. Then the insight of the decision-making process provides the different dimensions of the information.

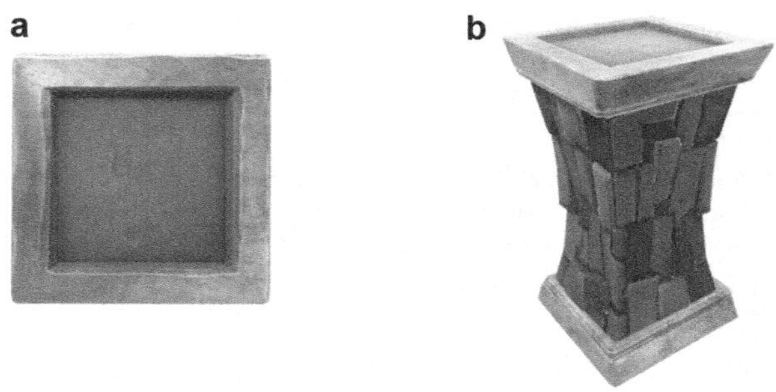

Fig. 7.1 (**a**) Top of the pillar, (**b**) Complete 3D pillar

To know more regarding the decision-making, we need to come across the below stages:

- Decisions are judgements
- Start the analysis with opinions over facts
- Know the criteria of relevant information/data
- Test your opinions against reality/actual data

7.3.1 *Decisions Are Judgements*

Decisions are judgments; it provides the alternatives of choices. This is not the choice between right or wrong. Instead of that, it provides the best of all choices between "almost right" and "probably wrong." To make the decisions, often people start with the facts. But executives who make effective decisions will not start with facts instead of that they will identify the opinions. Opinions are untested hypotheses but they must be tested against reality to identify effective decision-making. Many technologies are trying to improve the decision-making process in organizations, but it is very hectic to define and select the appropriate model for decisions.

7.3.2 *Know the Criteria of Relevant Information/Data*

Relevant information is data that helps to solve a problem. For any particular issue or scenario, the decision-makers need proper layout and the deeper level of detailed information can change and update the understandings of users regarding the future direction of a business. The only rigorous method is to test an opinion against reality. The decision authority of business preferences to add evidence that is relevant to the decisions of the communal, then it provides the performance of the entity.

7.3.3 *Test your Opinions Against Reality/Actual Data*

If the data is constructed with the reality and précised objective, the way of attributes are subjective and empirical, and Additional Veracities is nothing but a collective opinion. A familiarity in which some confidence is placed otherwise a rational collective illustration of "the way things are." Reality is not simply accredited, but must be discovered or coherent and is accountable to distortion.

7.4 Decision Intelligence: The Pathway

The pathway of decision intelligence provides actionable insights for customized business scenarios in the emerging artificial intelligence field.

- Expanding precision and recall value along with the decision intelligence techniques with ML and AI-based decision support.
- Intelligent integration with relevant data helps to develop predictive feasible outcomes.
- Enables the decision-making management systems with customized and homogenous also.
- Performance of data analytics leads to the openness of the decision intelligence pathway.

7.5 A Framework of DI

A framework of DI provides the traditional and advanced techniques to design the model. The execution of decision models and processes can be monitored and fine-tuned as per the dynamic business environment. When the businesses embracing the new principal or idea, it faces several challenges, especially the involvement to make the dynamic decision processes as decision intelligence. Even current business environment is willing to adopt the high-level guidance by the capabilities of the Decision Intelligence framework, but it does not know how to align a complex business problem as per the organizational objectives.

As such Fig. 7.2 shows a decision-centric approach is a critical component to any Decision Intelligence framework, as it will concentrate on a detailed strategy for successful implementation by identifying all of the relevant technologies and the attributes needed to meet a particular business requirement. Both data analyst leaders and AI enthusiasts are more focused on the accuracy of the algorithms and building them than understanding decisions, the impact and effectiveness them on business operations in the scope of their projects. The promise of the new discipline of "decision intelligence" is to provide a framework that brings all different aspects of decision-making into a unified view to addressing the above challenges.

7.6 The Emergence of Machines as Aids in Society

The proliferation of machines in human life in the past two centuries has revolutionized every aspect of our existence, e.g. the production process, the daily tasks in the home, treatment of patients in healthcare, etc. Since "the invention of the steam engine by Thomas Newcomen in the early eighteenth century," progress in technology has not stopped, but increased at a faster pace, especially in the

Fig. 7.2 Frame work of decision intelligence

past hundred years [1]. Improvements in Newcomen's invention revolutionized agriculture, transport and were a precursor of the industrial revolution. However, nothing would have predicted the advent of machines with human-like capabilities and intelligence, which has given rise to artificial intelligence (AI). Human beings are said to be intelligent because of their capacity for adaptation and improvisation in novel circumstances and non-routine tasks [2]. This human ability called intelligence is often said to be critical for innovation, efficiency, and productivity. Thus, humans were credited with intelligence unmatched by not other species on earth. But humanity's ingenuity is now giving rise to competitive species in the form of "intelligent machines" capable of performing tasks that traditionally only humans could perform. This has given rise to the term artificial intelligence. AI refers to technological systems that can replicate some of the behaviors credited to intelligence harbored by humans, e.g. reasoning, learning problem-solving, perceiving, improvising, etc. Basic forms of such artificial intelligence could be seen in machines such as printers that can remember saved features and print accordingly, personal computers which can store and recall information when prompted [3]. However, as technological advances increase, a more complex form of artificial intelligence has emerged that more closely mirrors human intelligent capabilities. Let us examine some of these more advanced forms of artificial intelligence and their capabilities.

7.7 The Evolving Pervasiveness of AI and Capabilities in Human Life

As explained earlier, modern machines created by humans are starting to exhibit significant abilities akin to the definition of intelligence. This means that they increasingly have the ability to perform more than the routine task. They venture into the complexity of planning, problem-solving, anticipating, and forecasting in some instances. This is the reason why they are increasingly used in fields that need the precision that human ability cannot attain, in a way those there are growing anxieties that AI represents a threat to humans [4]. It is not possible to capture all areas in which artificial intelligence is used in modern society given the diversity of contexts of its application [2]. In the next few paragraphs, we shall consider the application in the home, learning, transportation, healthcare, and business.

Almost everyone who lives in a city household nowadays comes across and uses AI nearly every day. Intelligent washing machines, for instance, can be programmed to do our laundry exactly as we like, with speed and efficiency. They reduce water usage and protect the fabric, preserving the clothes for longer. In the same perspective, modern cookers use AI to prepare household meals efficiently with minimum human intervention and labor. They require minimum attention and can cook without the risk of the food burning, spilling. Such devices include modern slow cookers. For home entertainment, modern television can record programs for humans to view later. Home security systems are trained to recognize our voices and fingerprints to give us access to our homes and keep intruders and trespassers away. These are significant developments that free man of man tedious and time-consuming tasks. It is also claimed that child cot death has significantly decreased due to intelligent devices such as baby monitors which alert parents when babies sleep in a certain position or stray off bed or cry, etc.

In learning, the advent of AI has also revolutionized inclusion and access, such as Google, Alexa, etc. A recent survey shows that 57 percent of people use these technological facilities multiple times a gay [5]. A major contributor to artificial intelligence's ability to aid in a wide variety of environments such as the educational environment, the home, and consumer activities, etc. is the access to almost infinite amounts of data on the internet [6]. This allows the AI to, at any point in time, given an internet connection, access more data than any one person could ever recall, in an instant. With this ability, the AI can answer a broad set of questions regarding almost any topic. This permits faster access to information than ever, reducing the time that one searches for answers which, in turn, frees time for more human learning in the same period. A deeper look can be taken into how this phenomenon gradually changes where and how consumers of all ages obtain the data they require. Children and young people in education can now easily with the likes of Google Assistant and others, gain insightful answers ranging from the translation of words and mathematical operations to brief accounts of historical events [5]. However, as described earlier the rapidly expanding roots of AI can acquire information on an unfathomable realm of data. Not only can artificial intelligence acquire this information for the user but more commonly it can now check the reliability of the source and the validity of such information. For many, there is no longer a need to even turn on the television in the morning, with one simple demand commercially available assistants will tell one everything they need to know to start their day including but not limited to the weather forecast, traffic conditions, their schedule, and any important notifications. AI is increasingly making everyday lives flow more fluidly and frees up time for humans, to do more human things.

An interesting place of implementation of artificial intelligence is in video games. Although this is not exactly a new implementation, its development and use have increased drastically in recent years due to the presence of intense competition. Video game companies, now more than ever, are competing to have the most complex in-game AI so that players of these games enjoy a very immersive experience. Advanced AI in modern games now can respond to a vast number of scenarios, player movements, interaction, and even interact with other AI characters in a multitude of ways. This is a clear demonstration of modern AI's ability to adapt to a wide range of scenarios, a skill previously unique to intelligent life. Furthermore, for the AI to be able to exhibit such behaviors and competencies, its decision-making skills must be fine-tuned and rigorously put through tests. The basis of many modern single-player games relies on decision-making by both the human and AI sides. Depending on the difficulty selected the AI characters—also referred to as Non-Player Characters (NPCs)—will have a different level of decision-making capability that it can utilize and in many games, the cores aspect of the game will be can the human playing the game make smarter and faster decisions than the AI to beat it [7]. AI has become complex to the point that it can challenge the very boundaries of human decision-making capacities. These properties of AI are having a growing presence in the industry. The next section shows some of the decision intelligence capabilities of AI in business.

7.8 AI, Decision Intelligence, and Business Organizations

In business, AI was traditionally perceived as a passive assistant which is deployed to do routine tasks to free up time for humans to focus on more strategic perspectives, thus supporting data-driven organization [8]. However, increasingly, AI is shaping its place in the strategic arena of organizations, showing its decision intelligence capabilities [6, 9]. This why Colson [8] suggests that business has to move towards a truly AI-driven practice. Distinguishing between data-driven entities and truly AI-driven ones is critical to position the business process in a technology-dominated competitive business environment.

References

1. A.B. Mahmoud, N. Grigoriou, L. Fuxman, D. Hack-Polay, F.B. Mahmoud, E. Yafi, S. Tehseen, Email is evil! Behavioural responses towards permission-based direct email marketing and gender differences. J. Res. Interact. Mark. **13**(2), 227–248 (2019). https://doi.org/10.1108/JRIM-09-2018-0112
2. F. Chollet, On the measure of intelligence, 2019. arXiv:1911.01547 [cs.AI], https://arxiv.org/pdf/1911.01547.pdf, Accessed 9 Jan 2020
3. A. Ghilan, A.P. Chiriac, L.E. Nita, A.G. Rusu, I. Neamtu, V.M. Chiriac, Trends in 3D printing processes for biomedical field: Opportunities and challenges. J. Polym. Environ. **28**(5), 1345–1367 (2020). https://doi.org/10.1007/s10924-020-01722-x
4. B. Marr, Is artificial intelligence (AI) a threat to humans? *Forbes*, 2020. https://www.forbes.com/sites/bernardmarr/2020/03/02/is-artificial-intelligence-ai-a-threat-to-humans/?sh=5a1078e5205d, Accessed 9 Jan 2020
5. Merritt, The impact of alexa and google home on consumer behavior. *Chatbot Magazine*, 2017. https://chatbotsmagazine.com/the-impact-of-alexa-and-google-home-on-consumer-behavior-c5753d838a38, Accessed 11 Jan 2021
6. S. Kuttappa, The rise of decision intelligence: AI that optimizes decision-making, 2020. https://www.ibm.com/blogs/journey-to-ai/2020/04/the-rise-of-decision-intelligence-ai-that-optimizes-decision-making/, Accessed 11 Jan 2021
7. M. Kopel, T. Hajas, Implementing AI for non-player characters in 3D video games, in *Intelligent Information and Database Systems*, ed. by N. Nguyen, D. Hoang, T. P. Hong, H. Pham, B. Trawiński, (Springer, Cham, 2018). https://doi.org/10.1007/978-3-319-75417-8_57
8. E. Colson, What AI-driven decision making looks like. *Harvard Business Review*, 2019. https://hbr.org/2019/07/what-ai-driven-decision-making-looks-like, Accessed 10 Jan 2021
9. L. Timms, What happens when AI meets HR? *Technology*, 2020. https://www.technologymagazine.com/ai/what-happens-when-ai-meets-hr, Accessed 10 Jan 2021

Chapter 8
A Survey on Analytics Technique Used for Business Intelligence

P. Dhivya, A. Karthikeyan, J. Ajayan, and S. Vigneshwaran

8.1 Introduction

Conventional BI is disturbed by stages that grow admittance to examination and convey higher business esteem. BI pioneers should follow how conservatives decipher their forward-looking item interests into a recharged energy and improved client experience. The BI and examination stage market are going through an essential move. During the previous 10 years, BI stage ventures have to a great extent been in IT-drove solidification and normalization anticipates for enormous scope frameworks of-record revealing. Presently, a more extensive scope of business clients are requesting admittance to intelligent styles of examination and experiences from cutting edge investigation, without expecting them to have IT or information science abilities. As the interest from business clients for unavoidable admittance to information disclosure abilities is developing, IT area needs to convey on this prerequisite without giving up governance. While the requirement for arrangement of-record answering to run organizations stays, there is a critical change in companies are fulfilling these and new business-client driven necessities. They are progressively moving from utilizing the introduced base, for example, customary and IT-driven stages that are the undertaking standard, to more decentralized information disclosure organizations that are currently spreading across endeavors. There is the change to stages that can be quickly actualized and can be utilized either by examiners and business clients to discover bits of knowledge rapidly or by IT to

P. Dhivya
Department of Computer Science and Engineering, Bannari Amman Institute of Technology, Erode, India

A. Karthikeyan (✉) · J. Ajayan · S. Vigneshwaran
Department of Electronics and Communication Engineering, SNS College of Technology, Coimbatore, India

P. M. Jeyanthi et al. (eds.), *Decision Intelligence Analytics and the Implementation of Strategic Business Management*, EAI/Springer Innovations in Communication and Computing, https://doi.org/10.1007/978-3-030-82763-2_8

rapidly construct investigation content to meet business necessities and to convey all the more ideal business benefits. Gartner gauges that in excess of a portion of net new buying is information revelation driven.

This move to a decentralized model, engaging more business clients, additionally drives the requirement for an administered information disclosure approach. This is a continuation of a six-year pattern, where the introduced base, IT-driven stages are being supplemented, and in 2014, they were progressively uprooted for new organizations and tasks with business-client driven information revelation and intelligent examination strategies. This is likewise expanding IT's interests and necessities around administration as arrangements develop. Making examination more available and unavoidable to a more extensive scope of clients and use cases is the essential objective of associations, making this transition. Traditional BI stage sellers have made a decent attempt to address the issues of the current market by conveying their own business-client driven information revelation abilities and luring reception through packaging and reconciliation with the remainder of their stack [1]. Notwithstanding, their contributions have been pale impersonations of the fruitful information revelation subject matter experts and, subsequently, have had restricted appropriation to date. Their interests in cutting edge information revelation capacities can possibly separate them and spike selection; however, these contributions are works in advancement. They are additionally consolidating further developed and prescriptive investigation worked from measurable capacities and calculations accessible inside the BI stage into examination applications. This will convey bits of knowledge to a more extensive scope of investigation clients that need progressed examination skills. As organizations actualize a more decentralized and bimodal administered information revelation way to deal with BI, business clients, and investigators additionally request admittance to self-administration abilities past information disclosure and intelligent perception of IT-curated information sources.

This incorporates admittance to complex, yet business-client open, information readiness devices. Business clients likewise search for simpler and quicker approaches to find pertinent examples and experiences in information. Accordingly, BI and investigation merchants present self-administration information readiness and shrewd information revelation and example identification capacities to address these arising necessities and to make separation on the lookout. The plan is to grow the utilization of investigation, especially understanding from cutting edge examination, to an expansive scope of purchasers and non-conventional BI clients, progressively on cell phones and sent in the cloud [2].

8.2 Literature Review

BI is completed through mining gadgets; these devices produce disclosures that are at last used to expand advantage over rivals. Mining instruments give better customers' relationship the heads CMR, through mining real penchants, plans, and even customers beat. Alpha customers are those that expect an essential part in a thing accomplishment, as such finding what they need is fundamental. This infers that, burrowing devices are central for stock advancing undertakings and publicizing associations [3].

Data mining approaches and instruments utilized for the business insight and stock examination. It portrays the utilization of data mining approaches in business insight (BI), which combined with information stockroom to utilize information mining innovation to give exact and the state-of-the-art data for compelling choice emotionally supportive network in business. The rundown of techniques utilized in the literature is dissected. This will help to give unavailable figures at the store/item level. The stock administration and gracefully chain the executives utilizing information mining methods will improve the business by giving successful interest examination, request estimating, and proper choice help for the business. This additionally gives the difficulties of those cycles regarding information size and unsure business conditions [4].

BI end clients should profit by specific norms of reports to get direction, structure, and even acknowledgment esteem inside the reports of the diverse movement fields. Also, a nonstop the norm meeting for all movement fields should happen in the administration bookkeeping division to make the current difficulties and center straightforward and organized in the division. This should be likewise reflected in the reports subsequently. At this data interface, a proficient and diagnostic update should be ensured to fit the organization's need. Developing normalized reports requires less exertion for information completions than in the past when using Microsoft Office items. The advancement is a one-time exertion and the further completion cycles can be controlled consequently [5].

8.3 Models and Techniques: An Overview

Business Intelligence uses key presentations markers for assessing present state for business blueprint. People are using web for every key need so data was opening up in more restricted stretches and have diminished deferments. The business utilizes remarkable worldwide bank related data. It depends on account of the automation and the usage of IT structures data.

Prescient demonstrating will envision esteems for a specific data thing for the specific characteristic. Bunching and oddity examination has comparable characteristics and is gathered as a set into classes. Exploratory data analysis is utilized to investigate dataset without a strong dependence on doubts. It is highlights recognize plans in an exploratory way. Online analytical processing, for example, OLAP instruments will engage a customer to examine assorted component of multi-dimensional information. Perception utilizing pictorial representation such as plots, histograms, and charts will be successfully seen Sequential and Temporal models examination Deviation, Trend, periodicity, Descriptive Data scattering, Correlation, Association and deep investigation connections between attributes are called as

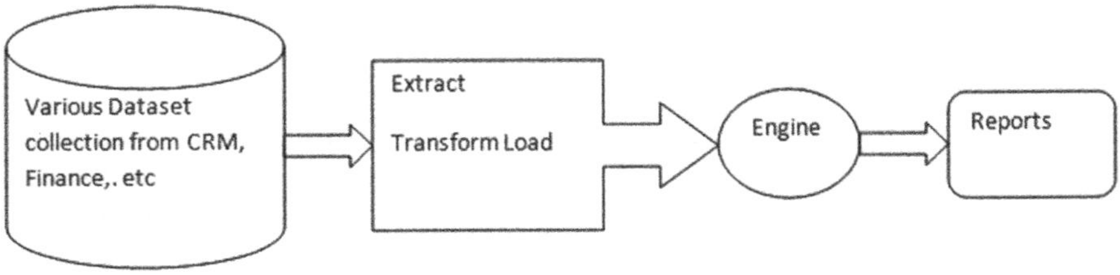

Fig. 8.1 Business intelligence analytics techniques

Identify Classification. It chooses to which class a data thing has a place [6]. The basic operations involved in BI analytics techniques were given in Fig. 8.1.

8.4 Applications

Inventory and demand determining are key business capacities for retailers and makers. Request determining encourages retailers to distinguish underachieving stores where the expected deals of an item seem to surpass the genuine deals of the item. Item distributions help makers to dispense items to stores and accounts. The primary aim of stock determining is to limit the inventory loading problem. A typical act of stock forecasting is to anticipate the interest for a specific thing in the future and hold the proper measure of things, in light of the forecasting results. Notwithstanding, stock information is a sort of time arrangement with enormous volume, long time frame, and less consistency. It is implementing an exact interpretable inventory forecast and modeling the connections among multiple time arrangement informational indexes and anticipating their future values simultaneously [7].

An advancement occasion anticipating framework called Promo Cast is accounted for, which utilizes a static cross-sectional relapse investigation of stock store deals under an assortment of advancement conditions, with store and chain explicit authentic execution data. The constraint of these examinations does not overlook the expected significance of value decreases and advancements of other persuasive items.

The two examinations increase the gauging exactness for retailers by incorporating special data from different items. Simultaneously, there are some significant contrasts. To start with, this paper considers both intra-and between classification limited time collaborations, while just considered intra-category competition [8].

Inconsistency recognition in time arrangement is one of the basic issues in information mining that tends to different issues in various spaces, for example, interruption discovery in PC organizations, anomaly identification in medical care tangible information and misrepresentation location in protection or securities. In any business, the investigation, stock administration, and successful choice emotionally supportive network are more significant. In stock administration, the inconsistency should be distinguished at the ideal time. It should give an ideal message or caution about the irregular stock information to the business administrator. In the paper creators built up a computational insight approach for value abuse recognition. The abnormality discovery on item costs is recognized utilizing concealed Markov models [9].

Vital BI advances and devices could be described by their capacity to essentially change the way in which business is conducted by furnishing a firm with a key preferred position. Business Intelligence technologies are seeing as instruments that support or change the undertaking's technique. The innovations in Business Intelligence are a framework that assists infrastructure with the different business systems. BI advances are ordinarily used to smooth out and increment the response time to ecological changes and to help the association in accomplishing a capacity. Key highlights of the key BI advances are choice emotionally supportive networks, undertaking asset arranging, information base frameworks, and ongoing data frameworks. In the accompanying sub-segments, the separation will be centered on procedure and operational BI innovations and apparatuses [5].

Predictive analytics is the part of information mining worried about the forecast of future probabilities and patterns. Prescient analytics can be utilized to consequently dissect a lot of information with various factors. It incorporates grouping, choice trees, market container examination, relapse displaying, neural nets, hereditary calculations, text mining, speculation testing, choice investigation, and others. For example, a master card organization could think about age, pay, acknowledge history, and other socioeconomics as indicators to decide a candidate's danger factor when giving a Visa. In prescient displaying, information is gathered, a measurable model is figured, forecasts are made, and the model is validated or reconsidered as extra information become accessible [5].

8.5 Conclusion

The present examination work talks and executes the view comparative with changes made in the innovation and business which are given explicit standards. Utilizing information as an apparatus, there has been a quick change in the idea of execution and making factorization of given model in a particular environment. At first, the issue of BI was distinguished development in BI advancement was introduced. The devices utilized in the advancement of BI were examined. Finally, certain methods and applications using business intelligence are discussed. The business intelligence idea is a multi-dimensional point in which there is no careful or worldwide origination of what business insight is. This subject is to give a point by point assessment of business knowledge definitions and significant insight ideas, for example, the substance of each key idea depicting how knowledge ideas are identified with one another.

References

1. P. Desikan, C. DeLong, S. Mane, K. Beemanapalli, K.-W. Hsu, P. Sriram, J. Srivastava, V. Venuturumilli, Web mining for business computing, in *Handbooks in Information Systems*, (Emerald Group Publishing Limited, Bingley, 2009)
2. A.-A.R. Al-Azmi, Data, text, and web mining for business intelligence: a survey. Int. J. Data Mining Knowl. Manage. Process **3**, 2 (2013)
3. U.-M. Mocanu, D. Litan, S. Olaru, A. Munteanu, Information systems in the knowledge based economy, in *WSEAS Transactions on Business and Economics*, vol. 7, (2010), p. 1
4. A. Venugopal, P. Jisha, Data mining techniques for business intelligence and its applications –A survey. Int. J. Adv. Res. Comp. Commun. Eng. **6**, 8 (2017)
5. G.W. Boland, J.H. Thrall, R. Duszak. Business intelligence, data mining, and future trends. Journal of the American College of Radiology **12**(1), 9–11 (2015)
6. N.V. Reddy, S. Avinash, C.S. Sai, Business intelligence—a survey. Int. J. Mech. Eng. Technol. **9**, 1 (2018)
7. Chaudhuri, Surajit, U. Dayal, V. Narasayya. An overview of business intelligence technology. Communications of the ACM **54**(8), 88–98 (2011)
8. A. Nikolopoulos, D. Papafotiou, N. Nikolopoulos, P. Grammelis, E. Kakaras, An advanced EMMS scheme for the prediction of drag coefficient under a 1.2 MW th CFBC isothermal flow—Part I: Numerical formulation. Chem. Eng. Sci. **65**(13), 4080–4088 (2010)
9. Y. Cao, Y. Li, S. Coleman, A. Belatreche, T.M. McGinnity, Adaptive hidden Markov model with anomaly states for price manipulation detection. IEEE Trans. Neural Netw. Learn. Syst. **26**(2), 318–330 (2015)

Chapter 9
Decision Intelligence Analytics: Making Decisions Through Data Pattern and Segmented Analytics

Bikram Pratim Bhuyan, Jung-Sup Um, T P Singh, and Tanupriya Choudhury

9.1 Introduction

Business intelligence is the current buzzword where data exploration is probed upon ranging from descriptive, diagnostic, predictive, and prescriptive analysis. Ranging from the simplest of questions from "what happened," "why something happened," "what is the chance of something happening," and "what action to execute if something happens"; decision analytics in business is considered to be one of the most important tools. The basic requirement for any analysis is "data." The data captured is then processed to perform the exploration discussed. In a traditional methodology, the data is assumed to be in normal distribution so that various hypotheses could be postulated out of it with the help of various tests. Although the central limit theorem used for the normalization of data is of much use when the data is huge and the assumption made from the huge population of data that the sample means tend to follow the normal distribution; the assumption

B. P. Bhuyan (✉)
Department of Informatics, School of Computer Science, University of Petroleum and Energy Studies (UPES), Bidholi, Dehradun, India
e-mail: bpbhuyan@ddn.upes.ac.in

J.-S. Um
Department of Geography, College of Social Sciences, Kyungpook National University, Daegu, South Korea

T. P. Singh
Department of Informatics, University of Petroleum and Energy Studies, Dehradun, Uttarakhand, India
e-mail: thipendra@gmail.com

T. Choudhury
School of CSE, University of Petroleum and Energy Studies, Dehradun, Uttarakhand, India
e-mail: tanupriya1986@gmail.com

P. M. Jeyanthi et al. (eds.), *Decision Intelligence Analytics and the Implementation of Strategic Business Management*, EAI/Springer Innovations in Communication and Computing, https://doi.org/10.1007/978-3-030-82763-2_9

fails when exponential distribution is seen in the data collected (for example, stock market returns). To amalgamate various distributions, we now propose an alternative technique to store and process data.

In statistics, when the data is collected from various individuals bearing some properties or attributes at the same period of time, it is regarded as cross-sectional data. When the data is taken for various individuals for a property at different intervals of time, it is termed as time-series data. Panel or longitudinal data can be seen as a mixed data having the properties of both cross-sectional and time-series data. Here data is collected for the individuals bearing the properties at different time intervals. Thus panel data is multidimensional data that is being observed over time [1]. Analysis of panel data is important in the fields of econometrics, epidemiology, and social behavioral studies [2, 3].

To perform analysis, two important models and their variants are generally used to model panel data. The classification of the models is termed as fixed effects model and the other as random effects model [4–6]. When an error is estimated in the regression model for the data, the error term can change stochastically (random effect) or non-stochastically (fixed effect) over the individuals or time. The Durbin–Wu–Hausman test [7] is often used as a discriminating analysis between the fixed and random effects model. The test is found out to be neither a necessary nor a sufficient metric [8]. The primary disadvantage in these models and their variants is that we do not have a rigid model for panel data analysis.

In this chapter, the author tries to model panel data in terms of a lattice. The lattice generation is common for various statistically different data, and then rules are formulated or mined from the lattice making the analysis common for all.

9.2 Basic Terminologies

9.2.1 Panel Data Analysis

Analysis of panel (longitudinal) data using regression materialized in the 1980s. It made a huge impact in the field of econometrics and was widely applied in social behavioral studies. The regression model intimating panel data tries to predict the target variable $\eta_{i,t}$ of object (individual) "i" at time "t." The model can be represented as

$$\eta_{i,t} = \omega_i + \sum_{k=1}^{s} \rho_{k,i} \chi_{k,i,t} + \varepsilon_{i,t}, \tag{9.1}$$

where $i = 1, 2, \ldots, N$ are the individuals at time $t = 1, 2, \ldots, T$ having property (explanatory) variable k. ω_i is the constant or intercept term for individual "i" and $\rho_{k,i}$ is the property variable coefficient. $\chi_{k,i,t}$ is the measure of the property variable "k" of individual "i" at time "t." $\varepsilon_{i,t}$ is the error term.

Panel data regression model can be broadly classified into three types: hybrid model (lacking the individual influence), variable intercept model (bearing individual influence), and varying coefficient model (stanching individual influence) [1].

In the models without individual influence, the intercept term ω_i is the same for all individuals. Thus in hybrid model, $\omega_i = \omega_j = \ldots = \omega$ also $\rho_{k,i} = \rho_{k,j} = \ldots = \rho$. In variable intercept model, $\omega_i \neq \omega_j$ and $\rho_{k,i} = \rho_{k,j} = \ldots = \rho$, which means that the individual influences only on the intercept term. In varying coefficient model, $\omega_i \neq \omega_j$ also $\rho_{k,i} \neq \rho_{k,j}$. Variable intercept model can further be sub-classified into fixed effects model and random effects model. The Hausman test [7] that is neither a necessary nor a sufficient metric [8] is generally used to differentiate between the usage of fixed effects model and random effects model.

9.2.2 Formal Concept Analysis

Formal concept analysis (FCA) was explored and researched in depth by Wille [9] and is adopted in multiple domains like linguistics, medicine, anthropology, psychology, computer sciences, industrial engineering, sociology, biology, and mathematics [10]. Formal concepts are spawned from a binary context and are collocated to form a lattice in a hierarchical order. Formal context can be defined by a triple tuple formed by three sets (G, M, I) where G represents the objects set or items, M the properties set, meanings, or attributes, and $I \subseteq G \times M$ shows a binary relation between G and M, i.e., $(a, b) \in I$ specifies that object a has the property (attribute) b. The following operators produce the Galois connection where $\alpha \subseteq G$ and $\beta \subseteq M$:

$$\alpha' = \{b \in M | (a, b) \in I, \forall a \in A\} \tag{9.2}$$

$$\beta' = \{a \in G | (a, b) \in I, \forall b \in B\}. \tag{9.3}$$

In simple terms, the Galois connection α' collects attributes set common to an objects set and similarly β' amalgamates the objects set to an attributes set.

A pair (A, B) defines a formal concept where $A \subseteq G$ and $B \subseteq M$, which follows the equalities $A' = B$ and $B' = A$. Further, $A'' = A$ and $B'' = B$ where $(.)''$ represents the closure operator. Implication rule (between two attributes sets X and Y) $X \to Y$ exists if $X' \subseteq Y'$. Implication follows the Armstrong rules:

$$\frac{}{X \to X}, \frac{X \to Y}{X \cup Z \to Y}, \frac{X \to Y, Y \cup Z \to V}{X \cup Z \to V}.$$

The set $\Delta(G, M, I) = \{(A, B) | A' = B \text{and} B' = A\}$ follows a partial order relation $\leq$ defined by- $(A_1, B_1) \leq (A_2, B_2)$ iff $A_1 \subseteq A_2$ and $B_2 \subseteq B_1$. $\Delta(G, M, I)$ spawns a complete lattice satisfying the $\leq$ relation. As the number of concepts could be very large (exponential), stability [11, 12] was introduced as an index to measure "useful" concepts. FCA can also be generalized to multidimensional pattern structures [13].

Table 9.1 Context: "Living Beings and Water" [14]

	a	b	c	d	e	f	g	h	i
1	ψ	ψ					ψ		
2	ψ	ψ					ψ	ψ	
3	ψ	ψ	ψ				ψ	ψ	
4	ψ		ψ				ψ	ψ	ψ
5	ψ	ψ		ψ		ψ			
6	ψ	ψ	ψ	ψ		ψ			
7	ψ		ψ	ψ	ψ				
8	ψ		ψ	ψ		ψ			

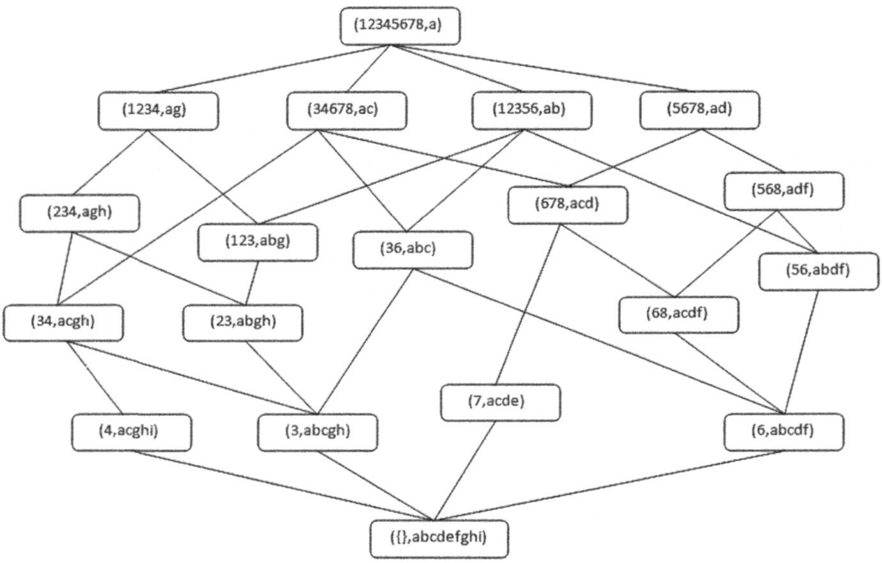

Fig. 9.1 Concept lattice: "Living Beings and Water"

Table 9.1 shows a context "Living Beings and Water" used in [14], and the concept lattice generated from it is shown in Fig. 9.1.

9.3 Formal Panel Concept Analysis

Here we give some definitions that introduce formal panel concept analysis.

Definition 1 A panel context Γ is defined as a quadruple (G, M, T, I), where G is the set of objects, M is the set of attributes, T is the set of time periods, and $I \subseteq G \times M \times T$. (g, m, t) ∈ I represents object g has attribute m at time t.

Definition 2 The derivative operators on A ⊆ G, B ⊆ M, C ⊆ T that produce the Galois connection is defined as

$$A^{'} = \{(m, t) \in (M \times T) | (g, m, t) \in I, \forall g \in A\} \tag{9.4}$$

(The set of attribute-time tuples common to the object set A)

$$B^{'} = \{(g, t) \in (G \times T) | (g, m, t) \in I, \forall m \in B\} \tag{9.5}$$

(The set of object-time tuples common to the attribute set B)

$$C^{'} = \{(g, m) \in (G \times M) | (g, m, t) \in I, \forall t \in C\} \tag{9.6}$$

(The set of object-attribute tuples common to the temporal set C).

Definition 3 The temporal derivative operators on A_t (object set A at time t) and B_t (attribute set B at time t) are defined as

$$A_t^{'} = \{m \in M | (g, m, t) \in I, \forall g \in G, \forall t \in T\} \tag{9.7}$$

$$B_t^{'} = \{g \in G | (g, m, t) \in I, \forall m \in M, \forall t \in T\}. \tag{9.8}$$

Definition 4 A panel concept of the panel context $\Gamma = (G, M, T, I)$ is defined as a triple (A, B, C), where A ⊆ G, B ⊆ M, and C ⊆ T, which satisfies the constraints $A_C^{'} = B_C$ and $B_C^{'} = A_C$. Also $A_C^{''} = A_C$ and $B_C^{''} = B_C$.

Definition 5 A panel concept (A_i, B_i, C_i) is said to be the super panel concept of (A_j, B_j, C_j) iff $A_j \subseteq A_i$, $B_i \subseteq B_j$ and $C_i \subseteq C_j$. (A_j, B_j, C_j) is said to be the subpanel concept of (A_i, B_i, C_i).

Definition 6 If $\Delta(A, B, C)$ is the set of all panel concepts generated from the panel context $\Gamma = (G, M, T, I)$, a partial ordered relation $\leq$ is then said to exist between all super panel concepts and subpanel concepts, which is denoted by $(A_j, B_j, C_j) \leq (A_i, B_i, C_i)$ iff $A_j \subseteq A_i$, $B_i \subseteq B_j$ and $C_i \subseteq C_j$.

Definition 7 $\Delta(A, B, C)$ w.r.t. the partially ordered property forms a complete lattice $< \Delta(A, B, C), \leq >$ with infima and suprema given by

$$\bigwedge_{h \in T} < (A_h, B_h, C_h) > = < \bigcup_{h \in T} A_h, \bigcap_{h \in T} B_h^{''}, \bigcap_{h \in T} C_h^{''} > \tag{9.9}$$

$$\bigvee_{h \in T} < (A_h, B_h, C_h) > = < \bigcap_{h \in T} A_h^{''}, \bigcup_{h \in T} B_h, \bigcup_{h \in T} C_h > . \tag{9.10}$$

Definition 8 Every panel concept produces some panel rules between the attributes of the concept at an instance of time. Formally, $\forall (A, B, C) \in \beta(A, B, C)$; $\exists B_i$ and

B_j, i, j $\in \{1, \ldots, m\}$ for each A, which follows $B_i \underset{C_t}{\Longrightarrow} B_j$, where $B_i, B_j \subseteq B$, $B_i \bigcup B_j = B$, $C_t \subseteq C$ and t $\in \{1, \ldots, t\}$.

Definition 9 The support σ of each panel rule $B_i \underset{C_t}{\Longrightarrow} B_j$ of the panel concept (A, B, C) is given as

$$\sigma(B_i \underset{C_t}{\Longrightarrow} B_j) = \frac{|\{B_i \bigcup_{C_t} B_j\}'|}{|M|} = \frac{|A_{C_t}|}{|M|}. \tag{9.11}$$

Definition 10 The confidence δ of each panel rule $B_i \underset{C_t}{\Longrightarrow} B_j$ of the panel concept (A, B, C) is given as

$$\delta(B_i \underset{C_t}{\Longrightarrow} B_j) = \frac{|\{B_i \bigcup_{C_t} B_j\}'|}{|B_i'|} = \frac{|A_{C_t}|}{|B_i'|}. \tag{9.12}$$

The above definitions help to first create and prepossess the data. Then a lattice could be generated with the panel concepts bearing the closure property. Finally, rules could be produced with a specific support and confidence level. The following section helps us to understand how the panel data is being represented.

9.4 Representation of Panel Data

We know that a data cube will be generated for the data collected. The cube being named as the panel context generated from panel data can either be represented as shown in Table 9.2 or as in Table 9.3.

Table 9.2 Attribute-object relation with time

	T_1				…	T_t			
	B_1	B_2	…	B_m	…	B_1	B_2	…	B_m
A_1	ψ	ψ					ψ		
A_2	ψ	ψ					ψ	ψ	
…	ψ	ψ	ψ				ψ	ψ	
A_n	ψ		ψ				ψ	ψ	ψ

Table 9.3 Temporal-object relation with attribute

	B_1				…	B_m			
	T_1	T_2	…	T_t	…	T_1	T_2	…	T_t
A_1	ψ	ψ					ψ		
A_2	ψ	ψ					ψ	ψ	
…	ψ	ψ	ψ				ψ	ψ	
A_n	ψ		ψ				ψ	ψ	ψ

We find some observations as follows:

Observation 1 Identical panel concepts are created from the context Γ_i in Table 9.2 and from the context Γ_j in Table 9.3. Also the counts are the same, i.e., $|\Delta(A_i, B_i, C_i)| = |\Delta(A_j, B_j, C_j)|$.

Observation 2 Isomorphic property is observed between the complete lattice constructed from the concepts in Table 9.2 with the lattice constructed from the context in Table 9.3.

Based on the above observation, we can create the context with respect to any of the architecture shown in Table 9.2 or 9.3. Now we can formally design our algorithms for data representation and rule mining.

Algorithm 1 Panel concept lattice

Construction of a lattice from panel-concepts of a given panel-context $\Gamma = (G, M, T, I)$
Input: Panel-Context
Output: Panel-Lattice

```
procedure CREATE—PANEL-CONCEPTS
    for each target attribute M_p ∈ M at time T_q ∈ T do
        Find M'_{pT_q} and M''_{pT_q}
        Form panel-concept (M'_{pT_q}, M''_{pT_q}, T_q)
    end for
end procedure
procedure CREATE—PANEL-LATTICE
    for each panel-concept (PC) do
        if ((M'_{xT_r} > M'_{yT_s}) && (M''_{xT_r} < M''_{yT_s}) && (T_r < T_s))
        Place (M'_{xT_r}, M''_{xT_r}, T_r) as the super panel-concept of (M'_{yT_s}, M''_{yT_s}, T_s)
        Draw an edge from (M'_{xT_r}, M''_{xT_r}, T_r) which is hierarchically above to (M'_{yT_s}, M''_{yT_s}, T_s).
    end for
end procedure
```

When forecasting is to be made on the panel data, it is made on the target attribute-time pair. Thus the context is generally divided into two groups. One is the target attribute-time pairs and other group is the non-target attribute-time pairs. We are interested in forecasting the target attribute-time pairs using the other. We now present Algorithm: 1 to build the lattice using target attribute-time pairs. Then Algorithm: 2 is used to generate rules for prediction.

Observation 3 The concepts created with the target attributes, using the closure property, will always result in the presence of target attributes in either the premises or conclusion of the associative rule.

Algorithm 2 Panel concept rules

Compute association rules from panel-concepts of a given panel-context
$\Gamma = (G, M, T, I)$
Input: Set of Panel-Concepts
Output: Set of Formal Panel-Rules

```
procedure CREATE—PANEL-RULES
    R = ϕ
    for each panel-concept (A, B, C) do
        if ((B_target ∈ B) && (C_target ∈ C))
        Produce rules R_i : B ⟹_C B_target
        R = R ⋃ R_i
    end for
end procedure
```

Table 9.4 Observations

Data set	Objects	Attributes	Time periods	Concepts	Rules
Labor market [15]	4165	120	7	38,413	21,510

9.5 Experimental Results

The data set used is "Cornwell and Rupert Labor Market Data" [15], and the question asked is whether wage (log wage) is related to education or other factors like experience, type of occupation, location, gender, etc.? Table 9.4 reflects the observations that were made after subjecting the data to our algorithms.

Common rules were generated like if $12 \geq$ weeks worked ≥ 17, log wage is between 2.0 and 2.5. If $20 \geq$ weeks worked ≥ 32, log wage is between 2.5 and 3.0. These sets of rules help us to identify the amount of work required to perform in order to receive the desired wage.

Interesting observations were made like people living in "south" having an experience of 17 years have increased their wages to 3 in their second year. People from south, working in manufacturing industry, got a log wage of 3.0–3.5 in their first year itself if he/she is not from manufacturing industry; he/she should be living in a city with a minimum experience of 5 years to get the same wage.

9.6 Conclusion

Panel data analysis is applicable in various fields. In this chapter, the authors tried to represent panel data formally in terms of a lattice, and then exploiting the properties of lattice, various rule-based analyses are being presented. Constitution of the data as a formal panel context is explored. The derivative operators produce a Galois connection that forms the panel concepts. The panel concepts then result in a complete lattice. Various rules are then generated using the lattice with support and confidence level. The algorithm is implemented in the data set "Cornwell and Rupert Labor Market Data" [15], and various interesting observations were generated.

References

1. B. Baltagi, *Econometric Analysis of Panel Data* (John Wiley & Sons, Hoboken, 2008)
2. E.W. Frees, *Longitudinal and Panel Data: Analysis and Applications in the Social Sciences* (Cambridge University Press, Cambridge, 2004)
3. C. Hsiao, *Analysis of Panel Data (No. 54)* (Cambridge University Press, Cambridge, 2014)
4. N.M. Laird, J.H. Ware, Random-effects models for longitudinal data. Biometrics 963–974 (1982)
5. G. Molenberghs, G. Verbeke, *Linear Mixed Models for Longitudinal Data* (Springer, New York, 2000)
6. J.C. Gardiner, Z. Luo, L.A. Roman, Fixed effects, random effects and GEE: what are the differences? Stat. Med. **28**(2), 221–239 (2009)
7. M. Nerlove, *Essays in Panel Data Econometrics* (Cambridge University Press, Cambridge, 2005)
8. T.S. Clark, D.A. Linzer, Should I use fixed or random effects. Polit. Sci. Res. Methods **3**(2), 399–408 (2015)
9. R. Wille, Restructuring lattice theory: an approach based on hierarchies of concepts, in *International Conference on Formal Concept Analysis* (Springer, Berlin, 2009), pp. 314–339
10. K.E. Wolff, A first course in formal concept analysis. SoftStat **93**, 429–438 (1993)
11. S.O. Kuznetsov, T. Makhalova, On interestingness measures of formal concepts. Inf. Sci. **442**, 202–219 (2018)
12. B.P. Bhuyan, A. Karmakar, S.M. Hazarika, Bounding stability in formal concept analysis, in *Advanced Computational and Communication Paradigms* (Springer, Singapore, 2018), pp. 545–552
13. B.P. Bhuyan, Relative similarity and stability in FCA pattern structures using game theory, in *2017 2nd International Conference on Communication Systems, Computing and IT Applications (CSCITA)* (IEEE, Piscataway, 2017), pp. 207–212
14. B. Ganter, R. Wille, *Formal Concept Analysis: Mathematical Foundations* (Springer Science & Business Media, Berlin, 2012)
15. C. Cornwell, P. Rupert, Efficient estimation with panel data: an empirical comparison of instrumental variables estimators. J. Appl. Eco. **3**(2), 149–155 (1988)

Chapter 10
Amalgamation of Business Intelligence with Corporate Strategic Management

Adham Kahlawi

10.1 Introduction

In recent years, the volumes of data available to companies are growing, and new technologies in business analysis are developing; consequently, the use of business intelligence tools is increasing.

Business Intelligence (BI) [1] means all the processes and tools through which a company manages to collect data of different nature to analyse them and draw strategic decisions.

In fact, business intelligence tools become a daily habit for some people even without knowing that they use it [2]. For example, the person who manages his/her company's account in LinkedIn can apply different analyses using the data collected from the interaction between users and the company profile. At the same time, the person who has an admin role for a Facebook page has access to a number of analyses that enable him/her to know the geographical distribution of the users who liked this page; in addition, can study the diversity of their opinions based on the diversity of their interaction with the different posts on the page, and the person who manages a hotel has the information of his customers including the personal data, the historical record of services and products requested by customers, period time and duration of customer reservations, and much more. These data can help predict the new strategy used by the hotel management to prepare for the upcoming season, including activities such as advertising and logistical preparations.

One of the advantages of business intelligence tools is their simple design that allows being used even by non-specialist users and understood easily by decision makers [3]. This happens thanks to BI dashboards that make it easy to integrate

A. Kahlawi (✉)
Department of Statistics, Computer Science, Applications, University of Florence, Florence, Italy
e-mail: adham.kahlawi@unifi.it

P. M. Jeyanthi et al. (eds.), *Decision Intelligence Analytics and the Implementation of Strategic Business Management*, EAI/Springer Innovations in Communication and Computing, https://doi.org/10.1007/978-3-030-82763-2_10

different data sources, easy to understand performance indicators with diagrams and graphs. Consequently, the dashboards allow even those who non-specialist in data analysis to have an instant idea of each type of activity in a few seconds; furthermore, it helps to prepare the reports easily in the form of visual dashboards.

Therefore, we can say the advancement of BI tools such as business analytics [4] makes the companies able to use the data collected to make strategic decisions faster and more effectively.

Business intelligence shows the business context data to help decision makers decide better strategic management [5]. Performance benchmarking helps the decision makers to make the control of the company more efficient and more manageable. Moreover, it facilitates identifying market trends, aimed at increasing sales or revenues, and it can also prove useful in compliance procedures and hiring. With business intelligence, you can optimise any aspect of strategic management.

10.2 Strategic Management

Strategic management is the way of leading a company based on a strategy. The traditional strategic management model [6] implies the belief that it is possible to formulate and develop a successful strategy through a rational process.

Strategic management is an integrated system of operations and activities related to analysing the internal and external environment, formulating an appropriate strategy, applying it and evaluating it in light of analysing the impact of environmental variables on it to ensure achieving a competitive strategic advantage for the company.

Companies have four levels of strategy, which are:

First, the organisation's strategy [7]It focuses on describing the company's overall and overall direction in terms of growth trends and methods of managing multiple activities. This level's fundamental responsibility is thinking about take advantage of the company's strengths and mitigate the impact of weaknesses in making strategic management such as the activity's type that the company has to participate in, the integration of joint projects, change the type of activity, the liquidation of activity, the distribution of resources, the flow of financial and non-financial resources to and from departments, between the company and interest groups, the entry points that the company can benefit from it in order to enhance profits and other tasks. The upper management layer provides the occasion for strategic unit's heads and major activities heads to improve the company's strategic vision. Strategies at this level are long term and have a widespread impact on the company as a whole.

Second, business strategy [8]

The business strategy usually takes place at the departmental level and focuses on improving its products or services' competitive position. It is stipulated that these strategies are consistent with the overall strategy of the company. These strategies attempt to give a high capacity for competition, innovation, market penetration, and opening new outlets for distribution. These strategies cover medium-term periods.

Third: the career strategy [9]

This strategy is drawn up planned by the business strategy and the overall strategy. This level identifies with the company's primary functions, such as production function, marketing strategy, finance plan, human resources mange, business administration function, and accounting departments. A manager responsible for one of these functional areas has to define the field's contribution to the implementation of the strategy. The strategies at this level are characterised by a short-term nature whose effect does not last for a long time.

Fourth, operational strategies [10]

Operational strategies are strategies for implementing career strategies. The issues here are urgent, fast, and require a robust, fast, and phased decision. The need for such a strategy appears due to the presence of emergency problems or tourist opportunities and does not bear the delay. Examples include the marketing strategies for facing specific threats from entering a new competitor, facing a promotional campaign, a deterioration in the quality of a commodity, a decrease in sales in a particular market, or a reduction in production in one of the production halls. The strategies in this part are strategies for implementing the project activities, representing quotidian behaviour or may cover a longer period but not more than 1 month.

The importance of strategic management can be summarised in the following points [11]. It improves the company's ability to deal with problems, makes the right decisions, reduces resistance to change, clarifies of future vision, achieves environmental interaction in the long term, strengthens the competitive position of the company, effective allocation of resources and capabilities; finally, it enhances performance and achieving satisfactory financial results.

10.3 The Role of Business Intelligence on Strategic Management Choices

Strategic management involves the formulation and implementation of the main objectives and initiatives taken by a top company's government [12], based on consideration of the resources and assessing the internal and external environments in which the company operates.

Strategic management is not static in nature; the feedback loop helps monitor the company objectives' execution and inform the next goals. Indeed, strategic management's dynamic nature provides the policies and plans to achieve its objectives, therefore allocating resources to implement the plans.

Determining strategic management has an important role; for instance, it helps to give your company a precise direction. It enables you to focus on where the business needs to grow, increase teamwork effectiveness, and collaborate within the team to measure business progress. It helps in a variety of other things.

To evaluate a company's current position before deciding on a new strategy, the decision makers can apply different analysis types as SWOT analysis [13]. Decision makers can use SWOT analysis to make the most of what you have for your company's benefit [14]. You can reduce your chances of failure by understanding what is missing and eliminating the dangers that would otherwise catch you off guard.

SWOT stands for Strengths, Weaknesses, Opportunities, and Threats, which means.

Strengths: Internal factors are indicators used by the company to apply immediate monitoring on the company activities; thus, it positively affects the company's business.

Weaknesses: All businesses, even the best ones, have some weaknesses. The absence of essential resources to contend in the market is weakness factors that lead to a disadvantage compared to the market. The essential resources can be in various fields, such as special skills, advanced technologies, lack of funding, creative marketing strategy and communication channels, and the absence of innovation.

Opportunities: Represent external factors that allow the company to benefit from an opportunity to achieve more economic profits, gain a better market position, or acquire a competitive advantage if adequately exploited.

Threats: Represent external risks factors that the company is exposed by the effect of the surrounding environment. In addition, the company cannot potentially exercise direct control over these factors; thus, it can only defend itself from prospective negative impacts.

The company's information collected and processed allows the evaluation and interpretation of the strengths and weaknesses of this company. Companies can influence strengths and weaknesses by monitoring, controlling, and analysing it through business intelligence software. The use of BI software for inventory management, accounting, customer intelligence and beyond, and others can increase the decision's quality, increase operational efficiency, and gain a competitive edge.

The company can improve its strengths and minimise as much as possible, its weaknesses by delivering the right information to the right people at the right time. In fact, that is what business intelligence is all about.

10.4 The Role of Data Quality in Strategic Management

Data quality [15] has always been an essential component for the company; indeed, it assumes a fundamental role for the company in an increasingly competitive market company and for its success in the businesses in which it operates.

The awareness of the data quality [16] has in supporting informed decisions and, conversely, of the disastrous consequences of inaccurate data. It has grown hand in hand with the spread of information sources available to companies, creating more substantial the need for adequate management for the company's data ever.

The standard ISO 8402 has defined the data quality as "The totality of characteristics of an entity that bear on its ability to satisfy stated or implied needs".

The previous definition tells the decision makers that the data quality does not depend only on the characteristics of the data itself but also on the business context in which it is used. Data quality is a critical component of the company [17]; for this reason, not implementing a strategy for evaluating and controlling the data quality is possessed can have disastrous effects. The presence of low data quality is not a theoretical problem but a real business problem [18]; consequently, it negatively affects the company's critical decisions.

The challenge of determining the company's competitive level in the market depends on strategic management; meanwhile, strategic management is affected by the quality of the data used from the decision makers at the time of its approving [19]. All in all, without data, strategic decisions are only hypotheses; on the other hand, the data and especially the high quality data makes every strategic decision, informed and targeted.

There are numerous traditional data sources such as relational databases, excel sheets, and social networks in each company. At this point, the problem becomes guaranteeing the quality of the data, primarily if used to support strategic management; in fact, guaranteeing the data quality is the critical factor of extreme importance in all business intelligence projects.

The main dimensions to measure the data quality in a given context are the following [20, 21]:

- Accessibility: it indicates the ease with which a user can identify, obtain, and use data. Business intelligence is interactive and approachable where users can customise dashboards and create reports on little notice.
- Accuracy: it refers to the difference between an estimate of how an attribute should be valued and the actual value reported by the data. Business intelligence tools can control this aspect of quality.
- Trustworthiness: It indicates the degree of credibility and reliability of the data, depending on the source of origin's reliability. Business intelligence platform appears data flaws very evidently once the company starts to use it.
- Completeness: it is a measure of correspondence between the real world and the specific dataset. Business intelligence tools indicate how much data is missing in the dataset to offer a 100% complete representation of the real context.
- Consistency: The degree of consistency of the data to obtain a consistent representation of the data. Business intelligence tools clean the data to ensure that the system's data is compatible to avoid garbage in/garbage out problems.
- Interpretability: it refers to the availability of clear and precise documentation of the database that indicates to decision makers which types of data are contained in the database and how to use and analyse it. Business intelligence methods use state of the art to create interactive dashboards that turn complex data into easily interpretable dashboards and graphs.

- Punctuality: it indicates how updated the data is concerning the real context. It is a measure of temporal alignment of the database with respect to the real world and constitutes an indicator of fundamental importance. Working on outdated data can lead to making bad critical decisions. Business intelligence uses analytics and other data processing tools to give companies access to the most recent data.
- Quantity: indicates how appropriate the volume of data held about a specific activity. Working with more or less data than necessary can be counterproductive and difficult to manage. Companies use business intelligence software to extract valuable insights from a specific quantity of data of their systems and operations.

10.5 The Business Intelligence for Development of Data Integration

Data integration is the process that defines the various steps: Data integration tools allow you to design these processes, one phase at a time and automate their execution. The most common data integration tools are known with ETL (extract, transform, load).

ETL (extract, transform, load) [22] is the acronym that refers to the process of extracting, transforming, and loading data from the most varied sources into a data warehouse or data mart for all business intelligence operations.

The role of an ETL process [23] is to feed a single, detailed, comprehensive, and high quality data source that can, in turn, provide a data warehouse. Very often the three operations of extraction, transformation, and loading are accompanied by a fourth cleaning operation, which is responsible for improving the quality of the data, avoiding the following situations: duplicate data, missing data, incorrect values, and values inconsistent.

The phases of the process are divided as follows:

Extract, the data is extracted from various sources, such as databases, activity logs, anomaly reports, security events, and other transactional activities;

Transform, the data then undergoes a transformation process that consists of selecting only those consistent with the system, eliminating duplicates, making connections between data from different sources, and aggregating them. Consequently, this transformation phase consolidated the data; in other words, this phase makes data coming from various sources homogeneous and making sure that the data adhere to the business logic of the analysis system for which it is developed;

Load, the last act of the entire data preparation process involves loading the data that extracted and transformed into a new destination. There are two distinct data replication models in the upload phase. In "push" replication, the application pushes the transformed data into the target database. In "pull" replication, the contrary, the target application or database requests the data, following the moment's specific needs.

Through the ETL process, the data acquire a high level of quality, so that it can be used for various operations:

- Data migration from one application to another;
- Replication of data for backup or redundancy analysis;
- Data entry into a data warehouse for assimilation, sorting, and transformation into business intelligence;
- Synchronisation of key systems.

In fact, ETL systems are business intelligence systems that considered as the key infrastructure for strategic management. This software allows you to transform unorganised data and content into strategically useful information to make critical decisions and operate effectively.

10.6 The Business Intelligence for the Development of a Reporting System

The report is presented as a combination of tables and graphs representing the relevant measures for the various phenomena analysed and disaggregated according to needs.

The reporting systems [24] are of strategic importance in keeping the company system in full efficiency and disseminating information to the various levels concerned. The reporting system is part of the programming and control systems. In fact, planning cannot be carried out if there is no information and data relating to the activities, resources used, and results obtained previously.

Reporting is fundamental and indispensable. No team will operate for long if operators have no information to provide feedback on their activities. With modern and efficient systems, they have to make people understand in time if the planned activities they are meeting expectations and if the set goals are achieved.

The different macroeconomic contexts of the company's activity create particular development environments; accordingly, a process of developing a reporting system is a dynamic process where its phases can be extended or reduced as a consequence of the particular development environments. Generally, composed of the following steps. First of all, its identification of information needs; then, its identification of the information context; after that, its identification of the information sources; next, its integration phase of information resources; then, its preparation of the report; after that, its identification of report visualisation needs; last of all, its validation of the report.

The decision maker of a company has to have a set of tools and techniques capable of providing functional mechanism strategic management. These tools represent a reporting system and the mensuration system of specific indicators. Consequently, all the information deems useful to understand better the business trend and support decision makers in making faster and more rational decisions.

Report generation is a standard application of business intelligence software. BI products can seamlessly generate reports for internal stakeholders and automate critical tasks for analysts. Thus, company managers can identify and modify corporate strategies and processes, supported by indicators provided in real time, which help guide choices, based on particular and measured elements.

On-Line Analytical Processing “OLAP” [25] indicates a set of BI techniques for the interactive analysis of large amounts of data, even in rather complex ways. The principle behind this technology is to search for data much more efficiently to allow analysts to make more complex queries aiming to develop the reporting system.

10.7 The Business Intelligence for Developing Future Scenarios

Scenario analysis is an essential tool in developing new strategies within the company [26]. This type of analysis explores firstly various paths that the company can follow then implement the one that best matches the company’s objectives.

Future analysis is a crucial tool for strategic management, as it creates future predictions and situations, analyses the environment, and prepares the company for possible scenarios.

Scenario planning is a model that can be used in strategic management to explore and learn the future of business strategy formation [27]. It works by describing a small number of scenarios, creating stories about how the future will evolve, and how this will affect specific issues affecting the market. The scenario planning method works by understanding the nature and effect of the most important forces affecting the future.

Predictive analysis systems are based on data collection and on the projection of reliable scenarios in the medium and long term, in order to provide indications and guidelines for strategic management. What-if analysis [28] is the necessary predictive analysis level based on data that can significantly contribute to making strategic management more effective, safe, and informed. What-if analysis in business intelligence scenarios indicates forecasting analysis techniques that allow evaluating a real system’s behaviour assuming a particular set of initial conditions. This type of analysis overcomes one of the fundamental limitations of reporting and OLAP, namely that of recording only the past and not allowing to analyse future scenarios.

10.8 The Business Intelligence for Optimising Processes

A company’s ability to face market needs competitively is closely linked to its ability to optimise its processes. The optimisation of processes and services in a company ensures this organisational efficiency and this ability to adapt; thus, continuously verifying the validity of the business processes implemented and the quality of the services provided [29].

Monitoring the processes helps identify the core business's critical points, improves its operations, and improves overall economic results. The optimisation of processes has become necessary for companies that identify cost reduction as the basis for maintaining their competitive capacity.

An optimal process can be defined as a set of activities that satisfy the client's expectations and all the company's technological and organisational constraints; besides, maintaining the primary goal of maximises profits and/or minimises costs.

The sequence of the main steps to be optimised can be summarised as follows:

- Identification and representation of existing processes
- Identification of the needs of the company
- Identification of critical processes
- Identification the roles and responsibilities of the organisational figures involved in the processes
- Identification of the systems of supporting the processes
- Identification and representation of the possible optimisation of processes
- Identification of any new roles and responsibilities of the organisational figures involved in the processes
- Identification of any changes to the information systems

BI tools are used to optimise a company's processes making a decisive contribution to reducing costs and increasing revenues.

In a fluctuating and articulated economic scenario like today, the need arises to interpret the market through an increasingly sophisticated analysis of the data in one's possession. BI is the primary tool for reading such information, which often lies unused in databases. Therefore, these data constitute a fundamental resource to photograph the business's health and define any corrective measures.

10.9 The Role of Business Intelligence on Obtaining a Competitive Advantage

Contemporary companies work in an external environment on a high degree of complexity and change; consequently, companies have to have methods and technical tools to study and analyse the market to prepare and implement the strategic management plan [23].

The study's importance and continuous environmental diagnosis of markets and competitors' position increase because of the expansion of markets, the multiplicity of competitors, the diversity of customer categories, the difference in their needs and desires, and their expectations. The importance of having clear developed tools and policies for conducting this study and that diagnosis is evident to monitor and analyse the opportunities, restrictions, or threats emerging from markets, customers, and competitors.

The term competitiveness [30] is extremely important in a world characterised by the speed of changes and their complexity in various fields as a result of the technological revolution. Companies cannot remain isolated from the static and non-static changes arising from multiple factors such as the communications revolution and global trade organisations' liberalisation. Companies have to pursue competitive paths and achieve Competitive advantages that guarantee its growth, survival, and continuity thanks to its competencies compared to its competitors.

Competitive advantage is anything that positively distinguishes a company or its products in a positive way from its competitors in the eyes of its customers or the end-users of its products. Furthermore, the competitive advantage arises from the value that the company is able to achieve for its customers and can be retained for a relatively long period due to the difficulty of imitation, as it can take in the form of low price or provide distinct benefits in the product compared to competitors.

Competitive advantage can be described as follows:

- it is relative
- it originates from within the company
- it reflects the efficiency of the company
- it affects the customer
- it works for the long term and does not go away quickly

The application of BI is considered a means to provide the organisation with the strategic information it needs to build competitive advantages that achieve a competitive advantage over competing organisations.

BI Analysis offers a company a competitive advantage in five aspects:

- Make faster decisions

  A regular and daily data collection activity allows management to examine the processes from a new perspective. Indeed, it will enable greater clarity on what is being done correctly. Meanwhile, the more advanced data analysis is the greater will be the competitive advantage over competitors in undertaking innovative business choices; especially, in those market sectors that are constantly evolving.
- Knowing the market

  Knowing the market is essential for making more efficient business decisions, and many BI tools can allow you to obtain in-depth information about your customers.
- Identify new market opportunities

  A company can obtain a competitive advantage with the help of data analysis. It can identify market opportunities not covered by other competitors and consistently assess the risks associated with pursuing them. In other words, a company will be able to develop new products of interest to the market before its competitors, thanks to knowing the trends and needs of customers by using BI tools that able to do so by relying on real data, not on assumptions of its management.

- Increase efficiency

  For BI to be useful, a company has to use it correctly, which is to understand how to simplify its processes and reduce costs.
- Take advantage of the benefits of being a small or medium company

  Data analysis can offer real-time indications on the need to change production or commercial processes, but it is difficult for large companies to reorient processes that have already started. On the other hand, a small or medium company can exploit their size to the advantage of greater flexibility and therefore change their strategies more quickly, according to the needs of the market.

10.10 Case Study

The year 2020 witnessed significant changes in consumer behaviour, especially after governments applying the lockdown. Indeed, many consumers started to count on buying their supplies through websites. Accordingly, the companies working in the field of E-Commerce found an excellent opportunity to increase their sales; however, this matter was not easy because the market witnessed the entry of many new competitors. Consequently, existing companies turn threatened to lose part of their market share. Therefore, these companies had to implement a particular strategy to increase their benefit from the external factors surrounding them and reduce the economic damage that could be inflicted on companies.

To implement this strategy, the Business Intelligence Department had to provide the company's management with information that enables it to understand its customers' nature based on the company's sales data. The historical data includes the following details, InvoiceNo, StockCode, Quantity, InvoceDate, UnitPrice, CustomerID, and CostomerCountry.

The Business Intelligence Department falsified the company's management with the following information:

1. Geographical distribution of customers

   Figure 10.1 shows that more than 80% of the customers live in England. In other words, the company's economic activity focuses almost entirely on customers residing in England, which indicates that it can expand the company into the markets of neighbouring countries to increase its market share.
2. Time distribution of purchases

   Figure 10.2 indicates that the purchases are at a similar rate during the first 9 months of the year, while this percentage increases in the last 3 months of the year.
3. Customer Segmentation

   It is not possible to deal with customers in the same manner; therefore, it is necessary to divide them into groups similar to their purchasing behaviour. To achieve this goal, we will apply the K-means algorithm depending on three variables that have been extracted from historical data, namely:

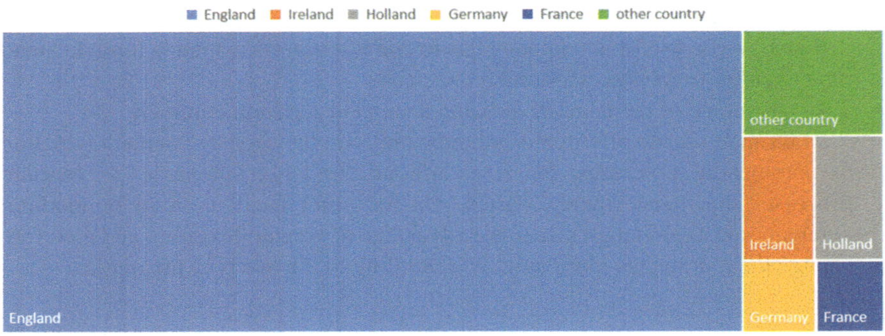

Fig. 10.1 Geographical distribution of customers

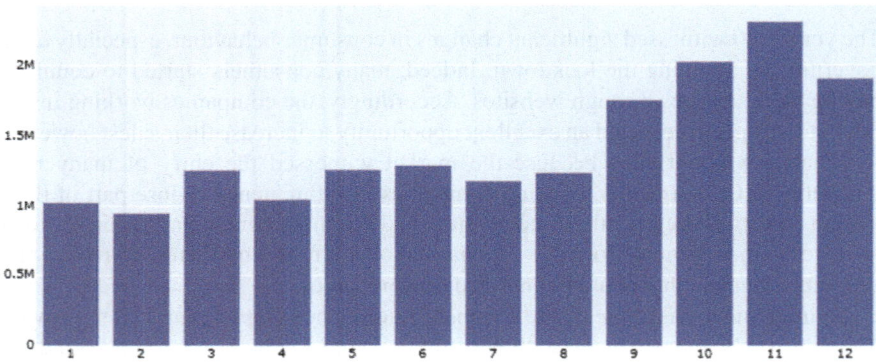

Fig. 10.2 Time distribution of purchases

Recency: represent the time since the last order.
Frequency: represent an average time between transactions.
Monetary: average transactions value.

Figure 10.3 shows that the customers can be divided into three groups. These three groups are characterised as the following:

Group 1: represent 48.9% of customers. Customers belonging to this group are distinguished by the fact that they make purchases frequently, but their orders range between a medium and a low value;

Group 2: represent 23.3% of customers. Customers belonging to this group are distinguished by the fact that they make purchases frequently; however, they do not make a new order for a long time, and their orders range between a medium and a low value.

Group 3: represent 27.7% of customers. Customers belonging to this group are distinguished by the fact that they make purchases infrequently, but their orders range between a medium and a high value.

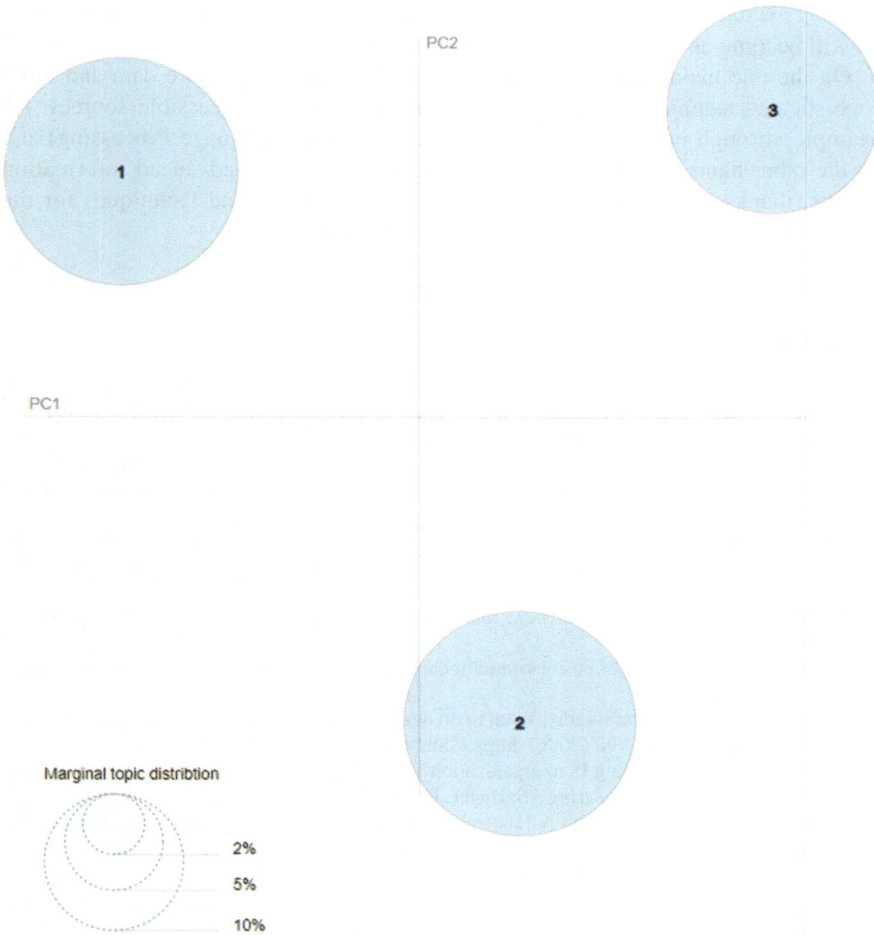

Fig. 10.3 K-means algorithm results

In conclusion, this report submitted by business intelligence will be the basis for the company's management, and in particular, the marketing department, to decide its future strategies.

10.11 Conclusion

To conclude this chapter, we can say, the future of Business Intelligence will be strongly governed by the technological advancement of Industry 4.0 and the growing availability and interconnection of data within companies. For this reason, BI will become an increasingly and evolved service.

On the one hand, it will be possible to take advantage of more data and more sophisticated techniques to extract data from previously inaccessible sources, for example, through natural language processing (Natural Language Processing) and on the other hand, it will be possible to create increasingly advanced information reports, thanks to the use of artificial intelligence models and techniques for the processing of insights starting from company databases.

References

1. X. Belle, Peng, Review of business intelligence through data analysis. BIJ **21**(2), 300–311 (2014). https://doi.org/10.1108/BIJ-08-2012-0050
2. R. Heath, Prediction machines: The simple economics of artificial intelligence. J. Inform. Technol. Case Appl. Res. **21**(3–4), 163–166 (2019). https://doi.org/10.1080/15228053.2019.1673511
3. Larson, Chang, A review and future direction of agile, business intelligence, analytics and data science. Int. J. Inform. Manage. **36**(5), 700–710 (2016). https://doi.org/10.1016/j.ijinfomgt.2016.04.013
4. Vercellis, *Business intelligence: data mining and optimization for decision making* (Wiley, London, 2009)
5. L. Arnott, Song, Patterns of business intelligence systems use in organizations. Decis. Support Syst. **97**, 58–68 (2017)
6. D. Wiltbank, Read, D. Sarasvathy, What to do next? The case for non-predictive strategy. Strat. Manage. J. **27**(10), 981–998 (2006). https://doi.org/10.1002/smj.555
7. Thevenet, Salinesi, Aligning IS to organization's strategy: The InStAl method BT, in *Advanced Information Systems Engineering*, (Springer, Berlin, 2007), pp. 203–217
8. M.-R. Bragge, Nurmi, Tanner, A repeatable e-collaboration process based on thinklets for multi-organization strategy development. Group Decis. Negot. **16**(4), 363–379 (2007). https://doi.org/10.1007/s10726-006-9055-5
9. F. Yean, K. Yahya, The influence of human resource management practices and career strategy on career satisfaction of insurance agents. Int. J. Business Soc. **14**(2), 193 (2013)
10. Tomlin, Wang, Operational strategies for managing supply chain disruption risk, in *The Handbook of Integrated Risk Management in Global Supply Chains*, (Wiley, Oxford, 2011), pp. 79–101. https://doi.org/10.1002/9781118115800
11. J. Teece, A capability theory of the firm: An economics and (strategic) management perspective. N. Z. Econ. Pap. **53**(1), 1–43 (2019). https://doi.org/10.1080/00779954.2017.1371208
12. P. Rony, Florinda, Knowledge management as a factor for the formulation and implementation of organization strategy. J. Knowl. Manage. **21**(2), 308–329 (2017). https://doi.org/10.1108/JKM-02-2016-0068
13. Gurel, Tat, SWOT analysis: A theoretical review. J. Int. Soc. Res. **10**(51), 51–66 (2017). https://doi.org/10.17719/jisr.2017.1832

14. M. Abdel-Basset, Smarandache, An extension of neutrosophic AHP–SWOT analysis for strategic planning and decision-making. Symmetry **10**, 4 (2018). https://doi.org/10.3390/sym10040116
15. C. Merino, S. Rivas, Piattini, A data quality in use model for big data. Futur. Gener. Comput. Syst. **63**, 123–130 (2016). https://doi.org/10.1016/j.future.2015.11.024
16. Kahlawi, An ontology driven ESCO LOD quality enhancement. Int. J. Adv. Comp. Sci. Appl. **11**, 60 (2020). https://doi.org/10.14569/IJACSA.2020.0110308
17. M. Smith, A. Roster, L. Golden, S. Albaum, A multi-group analysis of online survey respondent data quality: Comparing a regular USA consumer panel to MTurk samples. J. Bus. Res. **69**(8), 3139–3148 (2016). https://doi.org/10.1016/j.jbusres.2015.12.002
18. Choi, Luo, Data quality challenges for sustainable fashion supply chain operations in emerging markets: Roles of blockchain, government sponsors and environment taxes. Transp. Res. **131**, 139–152 (2019). https://doi.org/10.1016/j.tre.2019.09.019
19. R. Côrte-Real, Oliveira, Leveraging internet of things and big data analytics initiatives in European and American firms: Is data quality a way to extract business value? Inf. Manag. **57**(1), 103141 (2020). https://doi.org/10.1016/j.im.2019.01.003
20. G. Juddoo, Duquenoy, Windridge, Data governance in the health industry: investigating data quality dimensions within a big data context. Appl. Syst. Innov. **1**, 4 (2018). https://doi.org/10.3390/asi1040043
21. T. Taleb. A. El Kassabi. Serhani, Dssouli, and Bouhaddioui, Big data quality: A quality dimensions evaluation, in 2016 Intl IEEE Conferences on Ubiquitous Intelligence & Computing, Advanced and Trusted Computing, Scalable Computing and Communications, Cloud and Big Data Computing, Internet of People, and Smart World Congress (UIC/ATC/ScalCom/CBDCom/IoP/SmartWorld), 2016, pp. 759–765
22. Z. Fang, J. Elmore, A. Chien, UDP: A programmable accelerator for extract-transform-load workloads and more, in Proceedings of the 50th Annual IEEE/ACM International Symposium on Microarchitecture, 2017, pp. 55–68
23. A. Yulianto, Extract transform load (ETL) process in distributed database academic data warehouse. J. Comp. Sci. Inform. Technol. **4**(2), 61–68 (2019). https://doi.org/10.11591/aptikom.j.csit.36
24. E. Pearlson, S. Saunders, F. Galletta, *Managing and Using Information Systems: A Strategic Approach* (Wiley, New York, 2019)
25. S. Tohir, Kusrini, Sudarmawan, On-Line Analytic Processing (OLAP) modeling for graduation data presentation, in 2017 2nd International conferences on Information Technology, Information Systems and Electrical Engineering (ICITISEE), 2017, pp. 132–135
26. Quiceno et al., Scenario analysis for strategy design: A case study of the Colombian electricity industry. Energ. Strat. Rev. **23**, 57–68 (2019). https://doi.org/10.1016/j.esr.2018.12.009
27. Cairns, Wright, in *Evaluating the Effectiveness of Scenario Interventions Within Organizations BT-Scenario Thinking: Preparing Your Organization for the Future in an Unpredictable World*, ed. by G. Cairns, G. Wright, (Springer, Cham, 2018), pp. 247–255. https://doi.org/10.1007/978-3-319-49067-0_11
28. F. Hartmann, R. Moawad, L. Traon, GreyCat: Efficient what-if analytics for data in motion at scale. Inf. Syst. **83**, 101–117 (2019). https://doi.org/10.1016/j.is.2019.03.004
29. C. Carvalho, Cazarini, Gerolamo, Manufacturing in the fourth industrial revolution: A positive prospect in sustainable manufacturing. Proc. Manuf. **21**, 671–678 (2018). https://doi.org/10.1016/j.promfg.2018.02.170
30. Malik, Creating competitive advantage through source basic capital strategic humanity in the industrial age 4.0. Int. Res. J. Adv. Eng. Sci. **4**(1), 209–215 (2019)

Chapter 11
Role of Decision Intelligence in Strategic Business Planning

Syed Arslan Haider and Shehnaz Tehseen

11.1 Introduction

Decision intelligence (DI) is a fresh educational discipline anxious with all facets of choosing midst choices [1]. It carries a collection of the greatest pragmatic data sciences, communal sciences, and administrative sciences into an integrated area that assists a person's use of data to distinguish the lives, the industries, and the world around them. DI delivers facts that can add worth to the professional. With DI in place, industries can excerpt proofs from a great number of muddled figures at a very quick speed [2]. With immediate access to the professional data, we can examine central info and take more effective occupational conclusions. DI groups guarantee that the corporation becomes real-time innovative commercial reports to confirm that they can proficiently utilize the information for a healthier outcome. In an epoch of alphanumeric and mobile advertising, social media podiums, online movements, and market research; Aptitude aids as an decisive element in providing administrations and administrators to calculate their marketing outflows. DI can be interpreted into investigative intelligence where industries can take subtle conclusions on the foundation of figures, facts, and statistics, rather than conventions or perception [3]. DI with methodical abilities can offer all the arcade intuitions that can help to choose where to spend cash. Most prominently, by DI we can also recognize the smallest operative marketing policies so that we can eradicate those and amend the advertising budget.

Whereas, Strategic Business Planning (SBP) will mark sales tactics as more designed and operational [4]. Before tumbling the lead or arranging a commercial meet, it always supportive of the sale haze to recognize the leader's professionalism.

S. A. Haider (✉) · S. Tehseen
Sunway University Business School, Selangor, Malaysia

P. M. Jeyanthi et al. (eds.), *Decision Intelligence Analytics and the Implementation of Strategic Business Management*, EAI/Springer Innovations in Communication and Computing, https://doi.org/10.1007/978-3-030-82763-2_11

SBP can deliver all the data associated with the objective company's income, economical plans, upcoming extension plans, sales numbers, participants, and much more [5]. This exploration can give the organization auctions team more evidence to examine and prepare the sales arena on that origin. Deprived of having SBP, it will be continuously hard for sales directors to forecast which sales method can be used to change the lead or finishing the deal rapidly. Sales cannot occur on forecasts; it occurs over exploration and appropriate demonstration. Sales experts use SBP and make alterations in their tactic to formulating the sales arena more precisely. When an organization has all the figures, advertise visions, stats of purchaser buying ways, participant market policy; it is noticeable that the organization will exertion more on creating the corporate model and result-focused on. DI helps to examine the company's data and deliberate all the occupational material and strategic planning. SBP is essential for company evolution and achievement [6]. Similarly, SBP can deliver companies with the means to track progress, establish a modest outlook and make for unexpected variations in the marketplace. A deliberate plan contains many fundamentals a professional can utilize to fascinate financing and manage company purposes. To augment SBP, industries must clearly define corporation goals and demeanor widespread research to appropriately appreciate business trends.

11.2 Why Does Strategic Business Planning Need Decision Intelligence? And Decision Intelligence: How it Influences SBP?

11.2.1 Strategic Business Planning

The strategy is around solidifying the organization's exercises and utilizing and distributing constrained assets inside the organization to meet current objectives [7]. When planning a strategy it is vital to consider that choices are not taken out which any activity taken by the company may be met by the response of those included competitors, clients, workers or providers [8]. The procedure can also be characterized as the information of the competition, the vulnerability of events and considering the conceivable or genuine execution of others. A Strategic Business Planning (SBP) is a printed article that sets the purposes of a corporation with the requirements of the marketplace. Though a deliberate business plan comprises similar essentials of an outmoded plan, a strategic plan takes arrangement a step additional by not single essential corporation objectives but utilizing those areas to take benefit of accessible corporate occasions [9]. This is attained by cautiously examining a specific commercial industry and being trustworthy about the corporation's métier and meagerness in conference the necessities of the business. SBP is crucial to supplement market examination and to accomplish ideal arcade share for organization business [3]. The proposal allows industries to attention a specific forte in the marketplace, which creates sales and allows publicizing, and purchaser organization more effectively. The plan authorizations a business to know as much as the proposal allows industries to pay attention to the feasibility of the needs of its customers and fill in the cracks that are necessary for its longevity.

SBP helps an industry deliver healthier, more compact service to its clienteles. SBP embraces well-known market exploration, business dispositions, and contributor examines. The strategic plan will include the modules of an outdated plan, such as a policymaking summary, advertising investigation, and financial declarations, but a strategic plan will be additional accurate on how the corporation will go near about accomplishing company objectives. For instance, the SBP will struggle to recognize an objective arcade, then respond with a controllable market size, and create a strategy for gaining the customers [10]. Inscription an SBP has many rewards. The idea can assist as a summary of the fruitful conclusion of corporation signs. Company proprietors are in a healthier situation to not only appreciate their business but also become professionals in their productions. SBP assists administrators recognize the direction in which their formation is regulated by going through previous development and creating variations to improve and develop organization business. A plan is an administrative device that helps to maintain and keep an establishment on a path to achieve growth and financial goals. Many minor industries proprietors feel that strategic professional plans are for large establishments and big productions [11]. Though, Higgins et al. [12] conferring to the United State (U.S.) small trade management, a strategic business plan can revenue corporations of all magnitudes and can be an excessive benefit to small companies. Small industries may use the article to progress the policies necessary to attract and recall the clients its necessities to succeed.

11.2.2 Decision Intelligence

Decision intelligence (DI) is a manufacturing restraint that augments data science with philosophy from social science, resolution theory, and administrative science. Its solicitation provides an outline for best performs in managerial decision-making and procedures for relating machine learning at scale. Additionally, DI is the principle of whirling info into superior actions at any gauge. Facts are stunning, but its conclusions that are significant [13]. It's through our conclusions and our activities, that we aid in making a difference in the world around us. The word "decision" aids to the selection amongst variations of an object or element, therefore leading the discussion to wide-ranging choices than MBA-style quandaries (like whether to open a portion of association industry in London). It is due to our conclusions and movements that we distress the world round us. In this language, attaching a tag like that a cat against not-cat to a user's photo is a conclusion implemented by computer system, while calculation out whether to presentation that organization is a took decision unselfishly by the mortal leader (I hope!) in control of the project.

In our vernacular, a "Authoritative" is not that stockholder or depositor who falls in to consent the intrigues of the scheme group, but relatively the specific who is accountable for conclusion planning and situation mounting. In former arguments, a creator of meticulously-phrased purposes as opposed to their demolisher. "Authoritative" is best used inversely by dissimilar regulations, so it can be mentioned as taking an achievement when there were substitute choices. Execution aids to the humanly authority by adding a part of accountability and judgment [14]. A computer system though uniformly performs and implement a computer system can implement a conclusion, it not be entitled to a decision-maker because it does not tolerate accountability for its outputs that duty rests squarely on the shoulders of the entities who formed it. Additionally, not all productions/proposals are decisions. Indecision investigation terminology, a conclusion is only made once an irreversible allocation of resources takes place. By means of long as the organization can variation our minds for free, no conclusion has been made so far. Decision intelligence taxonomy is one way to learn a technique about conclusion aptitude which is interpreted along outmoded lines into its computable facets (largely overlying with practical data information) and partial facets (established mainly by intellectuals in the social and decision- making sciences). As far as the Partial side is concerned, the principles making up the partial side have conservatively been mentioned to as the conclusion sciences.

The *conclusion sciences* anxiety themselves with questions like:

- "How should organization set up judgment criteria and project metrics?" (All)
- "Is organization chosen metric impulse-compatible?" (Economics)
- "What excellence should an organization make this conclusion at and how much should organization pay for flawless information?" (Decision analysis)
- "How do sentiments, heuristics, and favoritisms play into decision-making?" (Psychology)
- "How do organic factors like cortisol levels affect decision-making?" (Neuroeconomics)
- "How will ever-changing the performance of information influence alternative behavior?" (Behavioral Economics)
- "How does one enhance organization results once generating choices during a bunch context?" (Experimental Game Theory)
- "How does one stability various restraints and time retro objectives in planning the choice context?" (Design)
- "Who can proficiency the inferences of the option and the way can varied teams comprehend that practice?" (UX Research)
- "Is the choice objective principled?" (Philosophy).

11.3 Decision Intelligence in Strategic Business Planning

Strategic Business Planning (SBP) is the progression of authenticating and launching a route of minor industry by seeing both where associations are and where organizations working [11]. The strategic plan stretches association an apartment to record job, image, and principles, and long-standing zones and the deed policies, association use to inspire them. A well-written strategic plan can play a crucial role in association minor industry's improvement and achievement because it tells association and employees how greatest to reply to occasions and happenstances. Nevertheless the profits of having a strategic plan in a dwelling, a growing number of small business proprietors are not joining on the long-term policies of their businesses. SBP consists of investigative the business and setting genuine goals and aims. This suggestion is to create an official document that lays out the organization's views and goals for the future. The SBP procedure can take certain time, but it is beneficial for everyone involved. As the minor industry owner, the association will have a healthier idea of the goals and aims, want to achieve and a track to prepare that. For association workers, the procedure can temporarily cause a growth in production therefore subsidizing the achievement of the industry [11]. The strategic planning process should contain association employees. Workers are elaborated in the day-to-day processes and be deliver a single view of the organization [15]. Workers can share concerning matters on what is and isn't functioning with the production on a daily basis, that can notify associations and prepare them for the upcoming constraints. In accumulation to workers, it is useful to touch out to individuals outside of the enterprise to acquire their views. Like workers, vendors have a unique standpoint on business. Talk to them about the industry, and get their opinions on how they think the industry backdrop can alter in the future.

Within a not-for-profit organization, DI and SBP will contain similar key modules. The DI and SBP may pay more attention to the inner and outer factors that can stand any pressures or trials of the association [13]. Because the construction of a charitable association can change quickly due to dissimilar aspects, the SBP receipts this into account and aims to address possible variations onward of time. Furthermore, the information technology (IT) business is continually altering. |It means a DI for an IT professional should identify and report the changes in the future and thinkable. Although other industry strategic plans may attention to the next 3–4 years, it is not rare for an IT strategic plan to gaze at the next year to year-and-a-half [16]. When it derives to evolving, studying, and apprising association IT planned a business plan, it is significant to include business's Chief Information Officer (CIO).

The U.S. small business management endorses that the SBP procedure is an elastic one [12]. Ever supposed why DI converted so much operational? The response maybe it is for the reason that of the revolt of e-commerce. Today whatsoever we need, it is just a tick away. For industry proprietors this means, they have to be very precise in inserting their produces, reviewing market trends, capitalizing at the right place for their advertising creativities, and appraising their efforts. DI helped them to shape a much more operational SBP quite than simply scrabbling around for the bright in the dim. DI can help companies recover their modest edge. DI is used to ascertain, share, and examine metrics and turn data into more tortious information while gain access to information about association competition. DI solutions unlocked the door to contact and reading modest intellect from one central data warehouse. This gives enchanted decision-making, quicker, and answering benefit to group industry consultants. Furthermore, involving association workers in the SBP procedure too means they accept a sense of answerability that can increase efficiency. Whether they contributed to the procedure or were up-to-date with the industry's goals and purposes after the strategic plan was created, they will be more likely to want to help the association attain those objectives. As part of the SBP process, we will scrutinize and examine the association's entire industry. We will take a look at what our professional does well and the zones where it still desires to recover. By recognizing the business's present powers and flaws, the procedure gives group and workers a chance to recover in the upcoming and come to be a sturdy industry by minimum dangers.

11.4 Decision Intelligence and Strategic Business Planning Misconceptions

There are many misapprehensions about SBP. Such as, it is time-consuming and it is only benefitting the larger businesses. Implementation of SBP may put us in the wrong direction that may not align with the business goals and objectives. For any business, small or large, it is beneficial equally if it is planned and implemented appropriately. It also helps the business to understand the position of the business and whether the managers' and employees' interests are aligned with the company's direction and objectives. Because it is difficult for a manager to judge if the company is going in the right direction. So, primarily, businesses need to consider the research phase of SBP. Final decisions and strategic planning should not be based on assumptions but should be based on research and valid information. This information may external or internal. Initially, the SBP process may be seen as scary but over time, it will be easy for the business, when managers better understand the process of how and what to do in it. It is time-consuming, but finally, it will better fit the concept of cost versus benefit analysis approach. It will be beneficial for the company to implement this process.

The managers' and employee's efforts can get a successful result from SBP. It also incorporates the contribution of the vendors and other external parties. If trained managers are involved in the planning of SBP, then the team members can better comprehend the SBP. Flexibility can also be considered during strategic planning. Because this is necessary for managers to accept any changes during implementation to achieve the aims and objectives of the business. It may result to achieve the goals later than expected and planned but it does not mean that the managers and the team are not ready to accept the change, having that the change will impact the performance of the business positively. On the success of the SBP, it is generally accepted that everyone in the team and organization is align with the goals and aims of the business. Everyone needs to contribute to the business's strategy to be implemented and they need to understand what to do and when to do it exactly.

11.5 Conclusion/Recommendations

Decision intelligence tools are used by various managers to know about what they want from the business but they do not consider that what they can do actually. DI helps the business to transform unstructured data into useful information. And the business can use this information to implement SBP, improve efficiency of the operations, and increase productivity. It offers the key perspective of customer leanings, the practice of buying, online shopping stats, etc. The enormous value that can be added to several businesses through this. While making strategic decisions, the company can get valued data resources through DI that can also help to achieve business goals. DI tools can also be used to interact with customers through emails, calls, and chats that will help to evaluate customer purchasing habits, market trends, etc. Eventually, the conversion ratio can also be increased through this data. The person must be concerned if he is CIO and responsible for the DI.

Schedule the meetings with your team and staff members who are involved in SBP continuously. Every meeting should have the valid agenda and a clear vision about the expectations of the performance of the business. It will help to keep the focus on the main purpose of the meeting and avoid disruption. Primarily, the meetings should be focused on the current performance of the business that the current position of your business, and the competitor's business. Comparative analysis can also be done with the competitor's data. When you got an idea about the current business then you can focus on the other analysis in upcoming meetings. Additionally, meetings with sellers, buyers, and other outside the business parties may also be held to collect data. The meetings with external parties may give you a more appropriate view of the situation about your business and your industry. External parties can give you deep insight into the factors that you may consider to determine the future achievement in business and industry.

Whenever you need to consider SBP (strategic business planning), and you think that it is necessary to start it urgently. Normally, you have to do it whenever you start any business but there is no specific rule to do SBP at the start of the business only but also you can do it during business any time [10]. If the business has started a long ago, then you can start SBP and DI too. It is never late to evaluate the present performance of the company and plan the strategy to obtain a position in the coming years. Any time, you can discuss and plan the SBP with your team. It is a continuous process. After the planning stage, it will lead to the development phase but the process is not finished yet. It needs to be implemented the plan and also monitor continuously to improve the business. If it is not planned realistically, then it is not possible to achieve the desired results. Owners of the business imagined the business to the maximum level but it is not good to consider the growth level over-optimistically. One thing to consider during SBP is commitment. If all the team is committed to SBP, then it is possible to achieve the required results. If the team is not focused on SBP, then the team will not align their interest with the company.

For this, SWOT (Strengths, Weaknesses, Opportunities, and Threats) analysis can be conducted. Through SWOT analysis, the performance of the business can be evaluated by you and your team. This will help to improve the performance of the business in the future because it will indicate any opportunity that business may take in the future and will lead to ultimate growth and success. It will also help to indicate any existing and new competitors that may prevent the success and growth of the business. SWOT analysis will help to identify business strengths that may use to improve the business. It also helps to know about the weaknesses of the business but it is not perceived as the negative impact of this model but also this feature will help to improve the business by mitigating the weaknesses. In this way, a business may consider the weaknesses in its future strategy. Opportunities can be easily identified as compared to threats but feedback from customers may ensure the business opportunities and threats also. Macroenvironment analysis can also be done to identify threats that may challenge the business in the future. In time identification of the threats and challenges are more important because if you know it before time then you can implement appropriate safeguards so that it may not affect the business negatively. When you have done all the data collection phase then you have to start the development phase by planning the business's strategies.

References

1. X. Zou, A.A. Scholer, E.T. Higgins, Risk preference: How decision maker's goal, current value state, and choice set work together. Psychol. Rev. **127**(1), 74 (2020)
2. B. Wieder, M.L. Ossimitz, The impact of business intelligence on the quality of decision making–a mediation model. Proc. Comp. Sci. **64**, 1163–1171 (2015)
3. J. Schmidtova, I. Formanek, Transformations in strategic business planning in the context of sustainability and business goals setting. Transform. Business Econ. **18**(2), 47 (2019)
4. B. Honig, M. Samuelsson, Business planning by intrapreneurs and entrepreneurs under environmental uncertainty and institutional pressure. Dent. Tech. **99**, 102124 (2020)
5. E. Elashkar, F. Aldeek, A. Shoukry, Business predictive analysis from business insurance data using business strategic planning techniques. Knowl. Manage. Res. Pract. **2020**, 1–10 (2020)
6. D. Rau, L. Flores, A. Simha, A moderated-mediated examination of the relations among strategic planning, organizational learning, slack and firm performance. Manag. Decis. (2020). https://doi.org/10.1108/MD-02-2020-0171
7. M. Reeves, C. Love, P. Tillmanns, Your strategy needs a strategy. Harv. Bus. Rev. **90**(9), 76–83 (2012)
8. D.C. Hambrick, J.W. Fredrickson, Are you sure you have a strategy? Acad. Manag. Perspect. **19**(4), 51–62 (2005)

9. G.F.G. Teixeira, O.C. Junior, How to make strategic planning for corporate sustainability? J. Clean. Prod. **230**, 1421–1431 (2019)
10. R. Rajnoha, J. Dobrovic, Managerial information support for strategic business performance management in industrial enterprises in Slovakia. Pol. J. Manag. Stud. **15**, 194 (2017)
11. R.I. Williams Jr., A. Smith, R. Aaron, J., C Manley, S., & C McDowell, W., Small business strategic management practices and performance: A configurational approach. Econ. Res. **33**, 1 (2020)
12. M.J. Higgins, D.J. Lacombe, B.S. Stenard, A.T. Young, Evaluating the effects of small business administration lending on growth. Small Bus. Econ. **2020**, 1–23 (2020)
13. T.T.T. Wee, The use of marketing research and intelligence in strategic planning: Key issues and future trends. Mark. Intell. Plan. **19**(4), 245–253 (2001)
14. F.J.B. De Oliveira, S. Ferson, R. Dyer, A collaborative decision support system framework for vertical farming business developments. Int. J. Decis. Support Syst. Technol. **13**(1), 34–66 (2020)
15. E.J. Bolland, Evolution of strategy: origin, planning, strategic planning and strategic management, in *Strategizing*, (Emerald Publishing Limited, Bingley, 2020), pp. 25–48. https://doi.org/10.1108/978-1-78973-697-720201003
16. A. Ilmudeen, Y. Bao, I.M. Alharbi, How does business-IT strategic alignment dimension impact on organizational performance measures. J. Enterp. Inf. Manag. **32**(3), 457–476 (2019)

Chapter 12
Social and Web Analytics: An Analytical Case Study on Twitter Data

Hitesh Kumar Sharma, Tanupriya Choudhury, and Hussain Falih Mahdi

12.1 Introduction

We are keen on how assumption can change by statistic gathering, news source, or geographic area. By extending our spatial examination of news elements to feeling maps, we can recognize topographical locales of good or unfavorable sentiments for given elements. We are additionally examining how much our opinion lists help us predicting future changes in society or market conduct. With the quick development of web-based life [1–5], sentiment analysis, additionally called opinion mining, has turned out to be a standout among the most dynamic research regions in natural language processing. Its application is additionally boundless, from business administrations to political campaigns. In our outcome, after characterizing the negative, neutral, and positive sentiments, we print the most regularly utilized words in our sentiment in the bar graph and pie-chart, helping the client to think about the sentiments more productively and successfully [6–8]. It is one of the speediest creating investigation zones in programming designing, making it an attempt to screen all of the activities in the area. Starting late, assumption examination has moved from exploring on the web thing studies to web-based life compositions from Twitter and Facebook. There are various types of reviews on social media platform

H. K. Sharma
School of Computer Science, University of Petroleum & Energy Studies (UPES), Energy Acres, Dehradun, India

T. Choudhury (✉)
School of CSE, University of Petroleum and Energy Studies, Dehradun, Uttarakhand, India
e-mail: tanupriya1986@gmail.com

H. F. Mahdi
Computer Engineering Department, Faculty of Engineering, University of Diyala, Baqubah, Iraq
e-mail: Hussain.mahdi@ieee.org

P. M. Jeyanthi et al. (eds.), *Decision Intelligence Analytics and the Implementation of Strategic Business Management*, EAI/Springer Innovations in Communication and Computing, https://doi.org/10.1007/978-3-030-82763-2_12

like product reviews, service reviews and comments, which is used for sentiment analysis based on social media data.

Talking about various fields' sentiment analysis or opinion mining is considered very important [9–13]. In publicizing field, associations go through it to fabricate their strategies, to grasp customer(s) feelings toward things or brand, how people respond to their items or things. In the political field, it is used to screen political view, to distinguish consistency and inconsistency among decrees and exercises at the organization level. In fact, election results are very well predicted by this analysis and if we think about it at the company level, this sentiment analysis will help manufacturer to estimate how well their product is doing in the market! This analysis is used to present the reality of social media emotions and identify the real mood of the public about the product or service, for the spotting of possibly dangerous conditions and choosing the general mien of the blogosphere [14–15]. End examination has been about supposition furthest point, i.e., paying little mind to whether someone has positive, impartial, or negative assumption toward something. Additionally, look at covering assessment examination and regular language getting ready has kept an eye on various issues that add to the congruity of end examination, for instance, ambiguity distinguishing proof, and multi-lingual help. Furthermore, concerning feelings, attempts are advancing from fundamental limit acknowledgment to extra complex nuances of emotions and isolating negative sentiments, for instance, shock and distress.

12.2 Social Media Platforms and Analytics Tools

12.2.1 Social Media Platforms

There are a plenty of social media platform active on internet but there are some important and old platforms which have a large active user base. A news float on these platforms impacts the business in positive and negative direction. The list of most popular social media [17–20] platform is given below (Table 12.1).

There are many factors that need to be considered before targeting a single or multiple social media platform. The target audience, the type of business, and the age group of audience are some of the important factors to be considered before collecting data from social media platform (Fig. 12.1).

In figure, we have shown some more statistics about seven most popular and widely used social media platforms. Figure is taken from [https://aofund.org/resource/choosing-right-social-media-platform-your-business/]. In this figure the purpose and the platform is best for ia also mentioned. These specifications will help to choose best suitable social media platform for collecting and analyzing social media data.

Table 12.1 List of most popular social media platform

Rank	Social media platform	Monthly active user (B = billion, M = million)
01	Facebook	2.23 B
02	YouTube	1.9 B
03	WhatsApp	1.5. B
04	Messenger	1.3 B
05	WeChat	1.06 B
06	Instagram	1.01 B
07	Tumblr	642 M
08	TikTok	500 M
09	Twitter	355 M
10	LinkedIn	294 M
11	Snapchat	255 M
12	Pinterest	250 M
13	Telegram	200 M

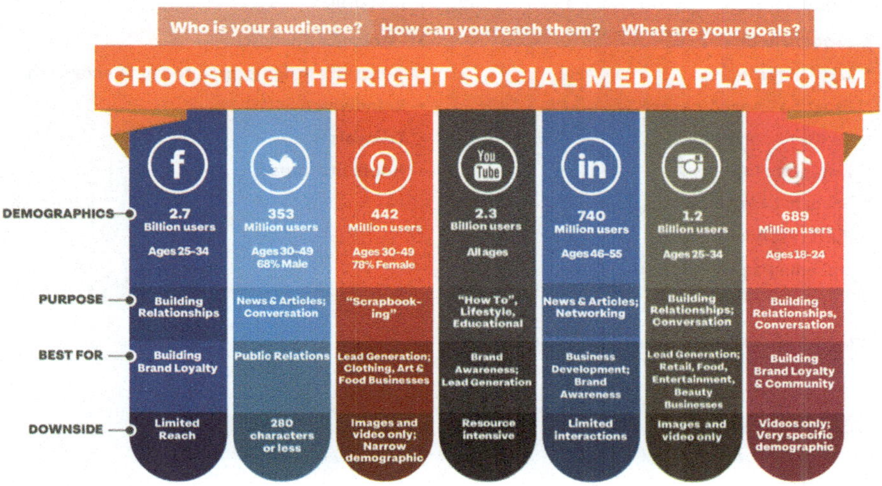

Fig. 12.1 Popular social media platform [https://aofund.org/resource/choosing-right-social-media-platform-your-business/]

12.2.2 Social Media Analytical Tools

Social media analytics or analytics of social media data analytics is a process of finding, collecting, and cleaning and analysis of data from social media platform. It helps the organization to keep track of public opinion about the product or service on important social media platform.

Social media analytics may follow two approaches:

1. *Single Social Platform Data Analysis*: In this analytics a single social media platform is focused by the analytical team. This is used for short term goals and finding instant feedback.
2. *Multiple Social Platform Data Analysis*: In this analytics a multiple social media platform is focused by the analytical team. This is used for long term goals and finding detailed feedback.

In Fig. 12.2, we have listed some popular tools used for social media analytics. We have given separate list for paid/free tools. In Fig. 12.3, we have shown an analytical dashboard of Google Analytics tool.

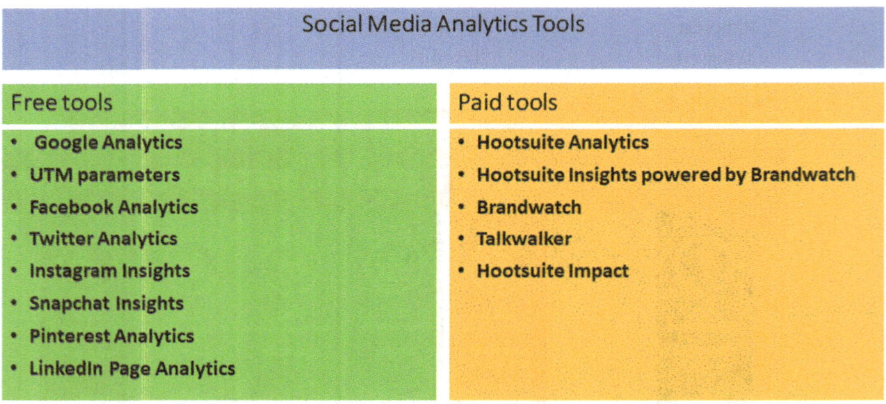

Fig. 12.2 Social media analytics tools

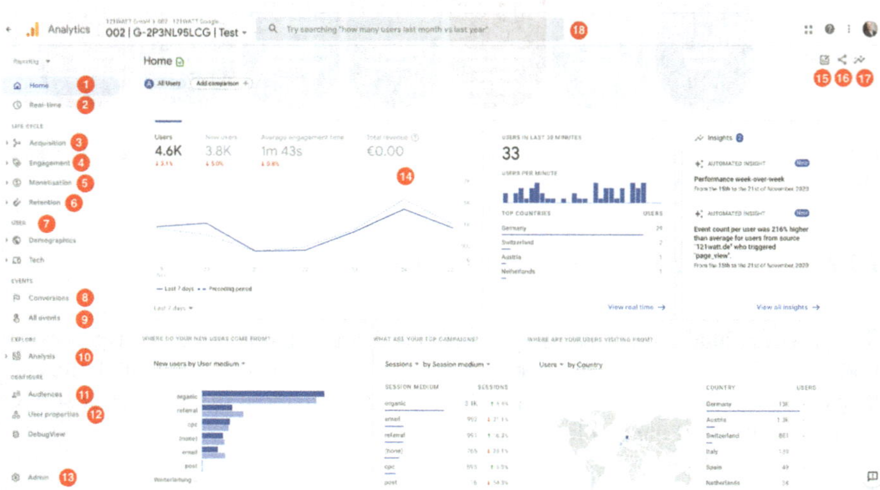

Fig. 12.3 Google analytics report screenshot [Ref. https://www.business2community.com/seo/google-analytics-4-properties-everything-you-should-know-02401869]

12.3 Social Media Data Collection Using Twitter API

There are many platforms for collecting dataset. Kaggel is one of the important historical dataset provider but for social media historical data is not feasible. Real time streaming of social media data needs to be captured. To make this process ease many social media platforms provide an API for fetching real time reviews from these platforms. Twitter provides Twitter API for integrating with various analytical platform. The step by step process of collecting and processing twitter data is given in following sections.

12.3.1 Data Collection from Twitter

Twitter REST API is used to collect tweets from Twitter. API key, API token, and API secret are provided to the authentic user to connect analytical platform with twitter live streaming.

In Fig. 12.4, the block diagrams have shown in which step by step process is explained.

In Fig. 12.5, a screenshot is given to display live feed fetched from twitter in R Studio.

Real time tweets are captured for analysis in R Studio.

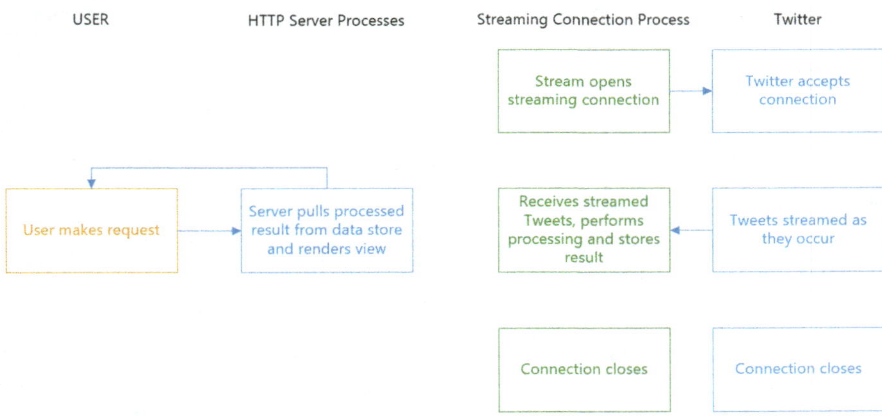

Fig. 12.4 Twitter API streaming flow chart for data collection request

12.3.2 Data Labeling

Data labeling is second step after collecting the data from social media platform. Mostly this task is done manually.

In Fig. 12.6, we have labeled the data with 0 or 1. This data is used for cyberbullying on social media platform, so tweets are labeled 1 if some abusing words are used and 0 if the tweet is normal tweet.

```
Forgoten_Indian b"RT @pbhushan1: Gujarat ranks 25/29 in infant mortality &amp; underweight children  with 39% malnourished. Thi
s is Modi's fabled Gujara "
NikhilGoyal46 b'  \nhttps://t.co/fRJnMxTm89'
rodadams46 b'RIDICULOUS #MAGA #Modi #auspol @younglabourUK https://t.co/CYhx54Be03'
aganpat b"RT @SurajPrSingh: PM @narendramodi got Sardara Singh in a semi hug and Sardara has his hand on Modi's shoulder..@Ra_T
HORe\n\nThis is th "
YogaSharer b'RT @MuscIeFitness: RT if yoga helps your mind, body, and soul! Check out the Free eBook below! \n#Namaste #yogi #y
ogalife https://t.co/NUS6P'
allahabadwali b'RT @allanmlobo: @BJP4India @narendramodi Modi ji personally checking if they have 1000 &amp;500 nots Our Indian
 Swamjiee are filthy rich look a'
reinergdrs b'@etzelgdrall direi di s, mi hanno soprannominato in tutti i modi, ma mai cos'
ranaalikash b'At a rally in Valsad district, Rahul Gandhi said PM Modi had failed to generate jobs for the people\n\nNana Pondh
a,  https://t.co/PHIuCIqhq9'
OfficeOfAS_ b'   !!! https://t.co/Zan8GmfSvu'
SultanK38669495 b'RT @HindiNews18:           \nhttps://t.co/8rJReeCnl0'
ParthkaSarthi b'RT @SushantBSinha:      8   550 , 23  .. 1400      https://t.co/T8l9NYxtFl'
rameshagrawal95 b'RT @narendramodi177: Gujarat Is Delighted To Warmly Welcome PM Modi ji, The Pride of Gujarat. #PMAtAkshardham
 https://t.co/oaLCVykihp'
Alikhansdaring b'#Happiest5WordSentence modi and company are prisoned'
siyawardas b'RT @TrueIndology: @thewire_in @rkarnad Notice the dirty tactics of leftists at work \n\nYogi Adityanath is respons
ible for oxygen thief '
allahabadwali b'RT @Saurabh83111542: @BJP4India @narendramodi Har har modi ji'
WithCongKerala b'RT @sidmtweets: If there was an election in UP today,anyone has any doubt that PM would have visited NTPC vict
ims rather than a temple prgr'
RajaShu83401000 b'@BJP4India @narendramodi @nsitharaman @PiyushGoyal V nyc modi g'
```

Fig. 12.5 Twitter API live tweet captured

	id	Insult	Date	Comment
0	1	0	20120603163526Z	"like this if you are a tribe fan"
1	2	1	20120531215447Z	"you're idiot......................."
2	3	1	20120823164228Z	"I am a woman Babs, and the only "war on women...
3	4	1	20120826010752Z	"WOW & YOU BENEFITTED SO MANY WINS THIS YEAR F...
4	5	1	20120602223825Z	"haha green me red you now loser whos winning ...
5	6	0	20120603202442Z	"\nMe and God both hate-faggots.\n\nWhat's the...
6	7	1	20120603163604Z	"Oh go kiss the of a goat....and you DUMMY...
7	8	0	20120602223902Z	"Not a chance Kid, you're wrong."
8	9	0	20120528064125Z	"On Some real Shit LIVE JASMIN!!!"
9	10	1	20120603071243Z	"ok but where the hell was it released?you all...

Fig. 12.6 Data labeling of Twitter tweets

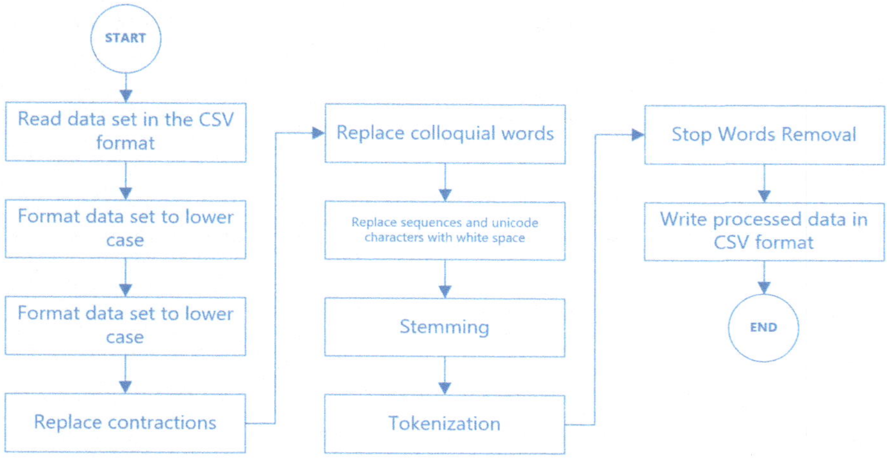

Fig. 12.7 Data pre-processing and labeling

12.3.3 Data Pre-Processing and Cleaning

Data pre-processing is the next step after labeling. The steps involved in pre-processing and cleaning are given in Fig. 12.7.

Conversion into similar caps and replacing colloquial words is initial step, then stop words are removed from tweets. After completing all pre-processing steps the clean data is again written to CSV file for analysis.

12.3.4 Data Analytics

This is the final step of social media data analytics. We have used python to write the code and implement data analytics algorithm. One of the screenshot of implementation is shown in Fig. 12.8. Some in build libraries are used for content analytics. Since data captured from twitter is purely in text format so dictionary and vocabulary are also imported.

This is the final step of development and implementation of social media analytics. This kind of implementation provides some insight provided from social media content.

```
#Bag of words
from sklearn.feature_extraction.text import CountVectorizer
bow_vectorizer = CountVectorizer(max_df=0.90, min_df=2, max_features=1000, stop_words='english')
# bag-of-words feature matrix
bow = bow_vectorizer.fit_transform(combi['tidy_tweet'])


from sklearn.linear_model import LogisticRegression
from sklearn.model_selection import train_test_split
from sklearn.metrics import f1_score

train_bow = bow[:31962,:]
test_bow = bow[31962:,:]

# splitting data into training and validation set
xtrain_bow, xvalid_bow, ytrain, yvalid = train_test_split(train_bow, train['label'], random_state=42, test_size=0.3)

lreg = LogisticRegression()
lreg.fit(xtrain_bow, ytrain) # training the model

prediction = lreg.predict_proba(xvalid_bow) # predicting on the validation set
prediction_int = prediction[:,1] >= 0.3 # if prediction is greater than or equal to 0.3 than 1 else 0
prediction_int = prediction_int.astype(np.int)

f1_score(yvalid, prediction_int) # calculating f1 score
```

Fig. 12.8 Data analytics implementation in Python screenshot

12.4 Conclusion

As we know that sentiment analysis is useful in certain cases like microblogging, the general perception of the personalities among the people, product review, advertisement impact on the people, etc. We can generate various graphs from the polarity, semantic value, etc. Both the Lexicon method and machine learning were implemented for this purpose with a focus on the Lexicon method. These graphs can be used to further predict the future possibilities. Utilizing hashtags to gather preparing information proved valuable. Then there are various machine learning algorithms such as Naïve Bayes classification, bag of words technique, random forest, etc. We have implemented "bag of the words" algorithm for text classification and prediction. Thus, the user can make a future strategy based on the predictions made. These predictions may not be always accurate but give as estimate about the accuracy of our project. More research is needed to determine and understand the satirical and the ironical type of statements which do not exactly mean what their literal translation is. Features from an existing sentiment lexicon were somewhat useful in conjunction with our requirements and are useful in various domains in future. We can expand this analysis to other platforms like Reddit, etc. more and more people are getting connected to the digital world.

References

1. P.K.D. Pramanik, S. Pal, M. Mukhopadhyay, Healthcare big data: A comprehensive overview, in *Intelligent Systems for Healthcare Management and Delivery*, ed. by N. Bouchemal, (IGI Global, Hershey, PA, 2019)
2. A. Shastri, R. Biswas, A framework for automated database tuning using dynamic SGA parameters and basic operating system utilities. Database Syst. J. **3**, 4 (2012)

3. K. Kshitiz, Shailendra, NLP and machine learning techniques for detecting insulting comments on social networking platforms. in International Conference on Advances in Computing and Communication Engineering (ICACCE), IEEE, 2018, pp. 265–272
4. I. Khanchi, E. Ahmed, H.K. Sharma, Automated framework for real-time sentiment analysis, in International Conference on Next Generation Computing Technologies (NGCT-2019)
5. I. Khanchi, N. Agarwal, P. Seth, P. Ahlawat, Real time activity logger: A user activity detection system. Int. J. Eng. Adv. Technol. **9**(1), 1991–1994 (2019)
6. A. Bhushan, V. Jain, T. Singh, K. Munjal, Model for automated database tuning framework. Int. J. Contr. Theory Appl. **9**(22), 347–357 (2016)
7. K. Kshitiz, H. Singh, P. Kukreja, Detecting hate speech and insults on social commentary using nlp and machine learning. Int. J. Eng. Technol. Sci. Res. **4**(12), 279–285 (2017)
8. P. Ahlawat, S.S. Biswas, Sensors based smart healthcare framework using internet of things (IoT). Int. J. Sci. Technol. Res. **9**(2), 1228–1234 (2020)
9. B. A. Thakkar, M. I. Hasan, M. A. Desai, Health care decision support system for swine flu prediction using Naive Bayes classifier, in 2010 International Conference on Advances in Recent Technologies in Communication and Computing, Kottayam, 2010, pp. 101–105
10. R. Vaishya, M. Javaid, I.H. Khan, A. Haleem, Artificial Intelligence (AI) applications for corona virus disease pandemic. Diabetes Metab. Syndr. Clin. Res. Rev. **14**(4), 337–339 (2020)
11. A. Shastri, R. Biswas, An algorithmic approach for auto-selection of resources to self-tune the database. IJRIT Int. J. Res. Inform. Technol. **1**, 9 (2013)
12. L. McClendon, N. Meghanathan, Using machine learning algorithms to analyze crime data. Mach. Learn. Appl. **2**(1), 1–12 (2015). https://doi.org/10.5121/mlaij.2015.2101
13. H. Singh, A. Jatain, A review on search based software engineering. IJRIT Int. J. Res. Inform. Technol. **2**, 4 (2014)
14. D. Antolos, Investigating factors associated with burglary crime analysis using logistic regression modeling. Dissertation, 2011
15. H.K. Sharma, A. Maheshwari, A. Singh, SGA: The focal point for automated database tuning. IJRIT Int. J. Res. Inform. Technol. **2**, 5 (2014)
16. D. Antolos, D. Liu, A. Ludu, D. Vincenzi, Burglary crime analysis using logistic regression, in *Human Interface and the Management of Information. Information and Interaction for Learning, Culture, Collaboration and Business*, ed. by S. Yamamoto, (Springer, Berlin, 2013). https://doi.org/10.1007/978-3-642-39226-9_60
17. S. Bansal, P. Kumar, S. Rawat, T. Choudhury, Analysis and impact of social media and it's privacy on big data. in 2018 International Conference on Advances in Computing and Communication Engineering (ICACCE), 2018, pp. 248–253
18. N. Singh, T. Sharma, A. Thakral, T. Choudhury, Detection of fake profile in online social networks using machine learning, in 2018 International Conference on Advances in Computing and Communication Engineering (ICACCE), 2018, pp. 231–234
19. S.S. Gupta, A. Thakral, T. Choudhury, Social media security analysis of threats and security measures, in 2018 International Conference on Advances in Computing and Communication Engineering (ICACCE), 2018, pp. 115–120
20. H. Kumar, S. Sharma, T. Choudhury, P. Kumar, Impact of Facebook's check-in feature on users of social networking sites, in *Computer Communication, Networking and Internet Security*, (Springer, Singapore, 2017), pp. 611–620

Chapter 13
People Analytics: Augmenting Horizon from Predictive Analytics to Prescriptive Analytics

Anurag Singh, Hardeep Singh, and Ashutosh Singh

13.1 Introduction

Business organization in modern times has several technological challenges. To meet these challenges in this technological era, companies need to think beyond the historical, short qualitative understandings of employee's performance and workforce behaviour for becoming more proactive to the demands of the industry. In a competitive market, it is important to utilize the employee's potentials to the fullest for the organization success. In such sort of business environment, human resource remains one of the distinct factors which can be used as a significant element for the potential growth of the organization by adding a great value with his skills set [1].

It is evident from many successful companies that companies are great because of their efficient employees but sometimes few employees add minimal value to their organization due to their reduce productivity. It is said that the one which is not measurable is not controllable. In long run ignoring or not able to measure productivity of such employees may bring disastrous results for the organization. Analytics in the context of people help companies to identify the human capital that shows lesser productivity at specific tasks [2]. This helps organization to take calculated decisions on the employees performing with low efficiency. Conversely companies having high turnover or attrition rate can utilize the methods of people analytics to ascertain the actual cause of 'intention to leave' from the end of the employees. This will help the organization to make better policies for retaining

A. Singh (✉) · H. Singh
DoGM, School of Business, University of Petroleum and Energy Studies (UPES), Dehradun, India

A. Singh
School of Liberal Studies, Amity University, Jaipur, India

P. M. Jeyanthi et al. (eds.), *Decision Intelligence Analytics and the Implementation of Strategic Business Management*, EAI/Springer Innovations in Communication and Computing, https://doi.org/10.1007/978-3-030-82763-2_13

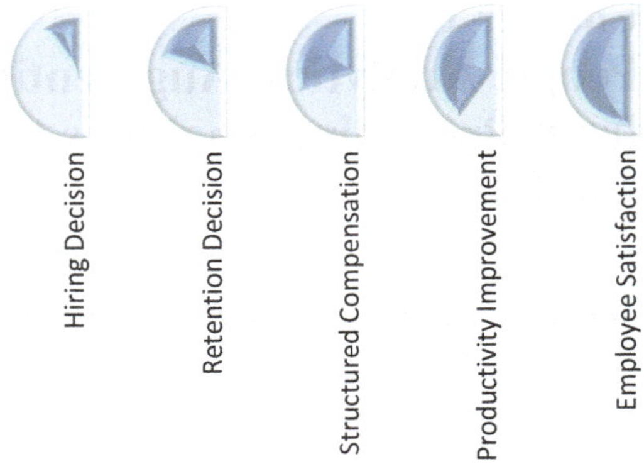

Fig. 13.1 Advantages of people analytics. Source: Authors compilation

the high performing employees. It also has an organizational impact, to aid in strategizing and better decision-making.

No matter how big a company is, finding right and skilful human resource is always a challenge for many organizations. HR decisions impact business performance [3, 4]. People analytics is really helpful in identifying the requirements of the organization based on the past data of a previous recruit, their performance and their success in the organization. Historical recruitment data helps companies to understand and make perfect recruitment plan for new hires. This helps them to analyse that whether a candidate is good fit for the company or not (Fig. 13.1).

Analytics is a scientific process of measuring, discovering and informing through meaningful patterns which can be found in the data. It focuses on converting unorganized data into insight for better decision-making. Analytics is heavily reliant on statistical analysis, computer programming and research. Davenport and Harris [5] define analytics as the systematic application of data, statistical and quantitative analysis, explanatory and predictive models and fact-based management to inform and guide decisions and actions [6]. People analytics, however, is a relatively unique intervention within the broader domain of human resource management. In a broader spectrum it is an element of decision sciences and data intelligence which uses statistical techniques and measurement tools for utilizing and concealing the effective decision-making such as workforce strategies and practices. It is often referred as workforce analytics or HR analytics ('people analytics'). People analytics can be elaborated as a data driven, objective oriented study of workforce wherein data and technology are used for the purpose of making factual decisions for the business and also for improving the business management [7]. Currently, the field is usually utilized in the human resources department of an organization. The process to perform people analytics is clearly given in Fig. 13.2.

Fig. 13.2 People analytics process. Source: Authors compilation

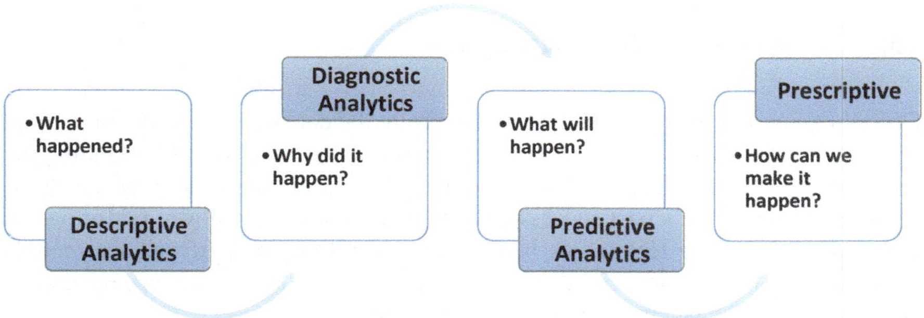

Fig. 13.3 People analytics constituents. Source: Authors compilation

13.2 People Analytics Constituents

People analytics is considered to be part of business analytics, the latter consists of three constituents that can be categorized as descriptive, diagnostic, predictive and prescriptive analytics. Firstly, descriptive analytics is a kind of data-based analysis that gives a detailed description of the overall performance of the organization and a high-level overview. Secondly, diagnostic analytics to identify the cause of issue [8]. Thirdly predictive analysis which can be defined as a technique used for predictive forecasting and helps in studying the recent and the historical facts and figures so that it could make predictions and see the near site of the future of the business and can predict unknown or unforeseen situations. Furthermore, prescriptive analysis, which may be thought of as contingency planning, identifies better methods to deal with particular situations via the use of both descriptive and predictive analytics and explains what might be the best option in a certain circumstance and how to deal with that specific case. Figure 13.3 presents the types of analytics used in people analytics.

As from the above discussion, it is evident that for important decision-making, predictive and prescriptive analytics can be considered as major constituent of people analytics. For people analytics, both predictive analysis and prescriptive analysis are equally important to study, as both of them help in improving decision-making and both improve the consequences of certain steps taken by the business.

It is clearly understood that we have to go through predictive analytics in order to reach prescriptive analytics. Predictive analysis uses modern techniques to evaluate data and helps to identify hidden patterns in the big employee data in order to prepare for future events. It uses segmentation and grouping techniques to analyse behavioural patterns and trends. Uses techniques such as grouping, rule-based and data-driven learning Once the process of predictive analytics is completed, we will transition to tactical analytics [9]. Prescriptive analytics then identifies the best response or action on the basis of predictions made by the predictive analytics.

13.3 Descriptive People Analytics

Descriptive people analytics analyse and measure the data of past events for gaining insight for future uncertainties. It evaluates past performance for understanding the performance of individual or group by mining historical data to look at the short-comings of employee's past success or failure. To analyse employee performance, we employ clustering techniques based on the similarity of performance metrics [10]. Almost every functional department, including sales, marketing, operations and finance, conducts a post-mortem review of this type.

Descriptive models quantify the relationships between data in a way that is frequently used to classify individuals or groups. It aids in the conversion of data into actionable insight for more informed and timely decision-making [11]. In contrast to predictive models, which are focused on predicting a single behaviour, such as turnover rates, descriptive models elucidate a variety of relationships between people and their behaviour. In contrast to predictive models, descriptive models do not rank individuals according to their likelihood of taking a particular action. Descriptive technical standards are being used to create additional models that can be used to evaluate workforce data in a number of ways. For instance, descriptive analytics analyses historical employee performance data to aid in the planning of training needs and enables businesses to conduct mock drills or placed higher training programmes.

13.4 Predictive People Analytics

Predictive analytics has become a critical component of people analytics. It transforms data into useful, relevant insights [12]. It analyses data in order to forecast the probable future outcome of an event or the probability of an occurrence. Predictive modelling is a statistical technique that utilizes a variety of statistical tools, including data mining, tree mapping, cluster analysis, text mining, trend analysis and game theory, to analyse current and historical data in order to forecast future events.

Modern business models employ predictive analytics to extract patterns from historical and transactional data in order to identify challenges and vulnerabilities. These models establish a relationship between the variables, enabling the threat or possible risk tied to a specific range of criteria to be evaluated, thereby guiding

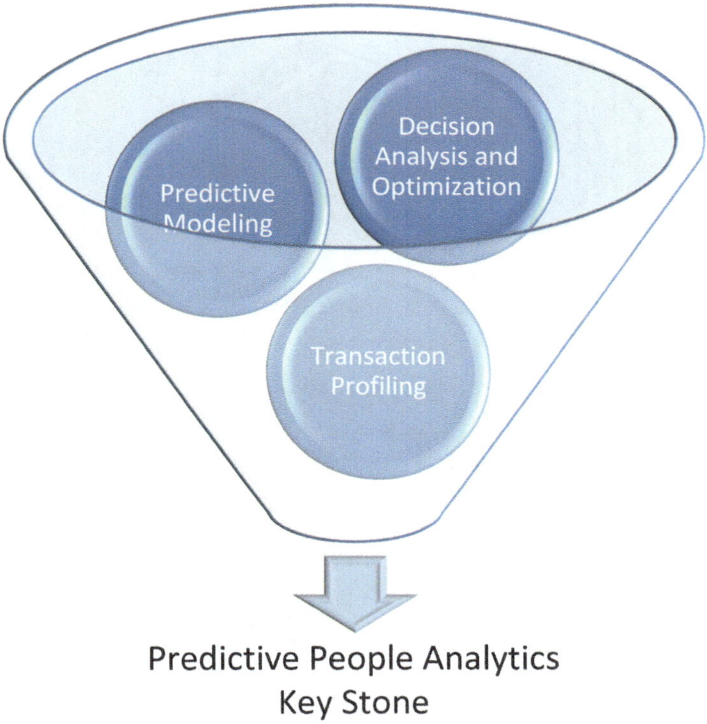

Fig. 13.4 Keystones of predictive people analytics. Source: Authors compilation

decision-making for candidate transactions. Figure 13.4 illustrates the three pillars of predictive people analytics.

Decision analysis and optimization are examples of how predictive analytics can be used to optimize employee relationship management systems (ERMS). It can assist organizations in analysing all employee data, thereby exposing patterns that can be used to forecast employee behaviour. Another example is for an organization that wants to analyse the performance of the employees, they can predict the performance on the basis of past data to observe trend and can take decision accordingly to find remedies in case of low performing employees. The efficiency of the current workforce is directly increased. For this reason, the company needs to invest in a squad of HR professionals and create a system based on statistical analysis to locate and access relevant data. The staff analysis team works with top management and all functional departments to develop a predictive information strategy.

A predictive analysis may be classified into six different categories, e.g. data mining, model identification and alerts, simulation of Monte Carlo, projection, analysis of root cause and predictive modelling (Fig. 13.5).

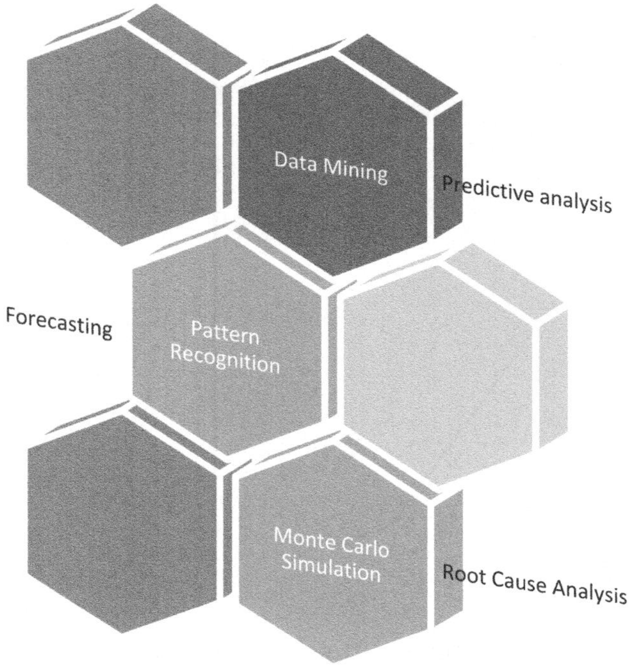

Fig. 13.5 Six categories of predictive analysis. Source: Authors compilation

1. Data mining identifies the relevant data related to the firm.
2. Pattern recognition allows us to take decision on taking measures of correcting any process.
3. Monte Carlo simulation talks about possibilities of action.
4. Forecasting enquires the 'what if' analysis in case the trend continues for a longer period of time.
5. Root cause analysis digs in deep to find the actual cause of a problem.
6. Predictive analysis talks about the further decision to be taken and what could be the next?

Predictive analysis bridges the gap from past understanding to present understanding and formulates its prediction accordingly to see the future event for the business [13]. The relevant statistical tool is used to check for the probability of the events that are about to happen in the near future by means of people modelling the prior data. It is a modern way of investigating scenarios and helps to recognize the major patterns from Big Data from the past and the present that help to predict or predict the future. Furthermore, prescriptive analysis can be divided into two parts:

1. *Optimization*: It asks about how to achieve the best option and what will be the best outcome.
2. *Stochastic Optimization*: It makes an enquiry about finding the best alternative to achieve the best result by avoiding complicacies.

13.5 Prescriptive People Analytics

Prescriptive analytics is a step beyond predictive analytics in that it goes beyond foreseeing upcoming scenarios and instead suggests measures to aid in accurate decision-making. Prescriptive analytics' implications are rooted in science and suggest the best alternatives. Prescriptive analytics anticipates not only what will occur and when it will occur, but also why it will occur. The output from descriptive and predictive measurement provides the foundation for prescriptive analytics to recommend directions for future based on the number of alternative solutions. Additionally, the latter can recommend alternative courses of action for capitalizing on future opportunities or mitigating future risks, as well as illustrate the implications of each decision option. In the real world, prescriptive analytics can be used to continuously and automatically process new data in order to make accurate predictions and provide improved decision-making options.

Prescriptive people analytics utilizes a variety of statistical and mathematical techniques. Prescriptive analytics data may come from a variety of sources, including internal (within the organization) and external, depending on the nature of the problem being studied. The data for the analysis may also be in structural form, which includes numerical and categorical data, as well as unstructured data, such as text, images, audio and video data, including big data.

Prescriptive analytics can aid strategic manpower planning by utilizing statistical methods to determine the surplus and shortage of employees within an organization, as well as the demand and supply of labour forces, in order to create accurate plans for filling the workforce gap within the organization [14]. Additionally, the decision will result in training, development, identification of possible resources via a potential appraisal system and decision-making regarding hiring and firing. Predictive analytics trends direct prescriptive analytics to take difficult decisions. This will only strengthen the organization by optimizing the operation of its most valuable resource, the human resource. Another example could be the establishment of the KRA's (Key Result Area) and KPI's (Key Performance Indicator) for the employees of the organization based on the results received from the prescriptive analytics. Prescriptive analytics can accurately predict many important decisions based on relevant manpower indicators. Also, it helps the organization to revisit the manpower policies for improvement.

13.6 Conclusion

Modern organizations strive to be predictive; they seek to gain insight from historical data in order to identify useful patterns and trends, forecast events, identify anomalies and evaluate changes in human behaviour in order to take suggestive measures that result in the desired business outcomes.

People analytics is a rapidly evolving area that many HR professionals find intimidating [15]. People analytics are widely used and significant [16].

It helps in enhancing employees experience and let the organizations make data-based decision, reduce employee attrition rate and improving employee productivity. With its varied method such as descriptive analytics, predictive analytics and prescriptive analytics, it serves every decision-making needs of the organization [17]. The decisions are made on the basis of historical as well as present workforce data. It can also be ascertained from the above discussion that predictive and prescriptive analytics are the major tools of people analytics and both considered to cater the important needs of the organization. Both of these analytic ways derived their existence from business analytics and are an advanced version designed according to the requirement of human resources decision [18]. Success in being predictive and proactive can be a game changer especially in human resource functions such as recruitment, attrition.

By focusing on both the entities we can conclude that prediction alone does not help us to undergo for best decision-making, prescriptive analytics plays a major role over there. Both are interlinked with each other, in different words we can say that both are dependent on each other to act effectively. It is not possible to jump on the rear part without evaluating the initial part.

References

1. Bharti, Human resource analytics. South Asian J. Market. Manage. Res. **7**(5), 68–77 (2017)
2. M. DiClaudio, People analytics and the rise of HR: how data, analytics and emerging technology can transform human resources (HR) into a profit center. Strat. HR Rev. **18**(2), 42–46 (2019). https://doi.org/10.1108/SHR-11-2018-0096
3. T.M. Qureshi, HR analytics, fad or fashion for organizational sustainability, in *Sustainable development and social responsibility*, ed. by M. Mateev, J. Nightingale, (Springer, Cham, 2020)
4. A. Boakye et al., The rise of HR analytics: Exploring its implications from a developing country perspective. J. Human Resour. Manage. **8**(3), 181–189 (2020). https://doi.org/10.11648/j.jhrm.20200803.19
5. T.H. Davenport, J.G. Harris, *Competing on Analytics: The New Science of Winning* (Harvard Business Review, Brighton, 2006)
6. A. Levenson, Using targeted analytics to improve talent decisions. People Strategy **34**(2), 34 (2011)
7. Nijjer, S. Raj, *Predictive Analytics in Human Resource Management: A Hands-on Approach*, 1st edn. (Routledge, New York, 2021)

8. D. Frazzetto et al., Prescriptive analytics: A survey of emerging trends and technologies. VLDB J. **28**, 575–595 (2019). https://doi.org/10.1007/s00778-019-00539-y
9. P.S. Deshpande, S.K. Peddoju, Predictive and prescriptive analytics in big-data era, in *Security and Data Storage Aspect in Cloud Computing*, (Springer, Singapore, 2019)
10. V. Kakulapati, Kalluri, et al., Predictive analytics of HR—A machine learning approach. J. Statistics Manage. Syst. **23**(6), 959–969 (2020). https://doi.org/10.1080/09720510.2020.1799497
11. D. Delen, S. Ram, Research challenges and opportunities in business analytics. J. Business Anal. **1**(1), 2–12 (2018). https://doi.org/10.1080/2573234X.2018.1507324
12. Q. Abdul, A. Mohammed, HR analytics: A modern tool in HR for predictive decision making. J. Manage. **6**, 51–63 (2019). https://doi.org/10.34218/JOM.6.3.2019.007
13. S. Malisetty et al., Predictive analytics in HR management. Indian J. Public Health Res. Develop. **8**, 115 (2017). https://doi.org/10.5958/0976-5506.2017.00171.1
14. S. Zeidan, N. Itani, HR analytics and organizational effectiveness. Int. J. Emer. Technol. **11**, 2 (2020)
15. F. Jamison, People analytics in practice: connecting employee, customer, and operational data to create evidence-based decision making, in *Smart service systems, operations management, and analytics*, ed. by H. Yang, R. Qiu, W. Chen, (Springer, Cham, 2019)
16. T. Peeters et al., People analytics effectiveness: Developing a framework. J. Organiz. Effective. People Perform. **7**(2), 203–219 (2020). https://doi.org/10.1108/JOEPP-04-2020-0071
17. J.W. Boudreau, E. Lawler III, What makes HR a strategic partner? People Strategy **32**(1), 14–22 (2009)
18. D. Angrave et al., HR and analytics: Why HR are set to fail the big data challenge. Hum. Resour. Manag. J. **26**(1), 1–11 (2016)

Chapter 14
Machine Learning Based Predictive Analytics: A Use Case in Insurance Sector

Hitesh Kumar Sharma, Tanupriya Choudhury, and Teoh Teik Toe

14.1 Introduction

An organization having large dataset can start data analytics using appropriate statistical methods and machine learning algorithms and fulfil its business informative goals. With appropriate knowledge of analytical approach and machine learning [1–4] algorithms the decision can be more favourable for business growth. Based on the business goal, data analytics can be categorized into four types as shown in Fig. 14.1.

14.1.1 Descriptive Analytics

It is the type of analytics in which big data is used to get insight from past data. The historical data is used to analyse the past records to find that answers of the related to what happened in past in the organization [5–7]. The past sales, revenues, feedback and process related data are analysed to get some useful information about that past and try to learn from these patterns.

H. K. Sharma
School of Computer Science, University of Petroleum & Energy Studies (UPES), Energy Acres, Dehradun, India

T. Choudhury (✉)
School of CSE, University of Petroleum and Energy Studies, Dehradun, Uttarakhand, India
e-mail: tanupriya1986@gmail.com

T. T. Toe
Division of Information Technology & Operations Management, College of Business (Nanyang Business School), NTU, Singapore, Singapore
e-mail: ttteoh@ntu.edu.sg

P. M. Jeyanthi et al. (eds.), *Decision Intelligence Analytics and the Implementation of Strategic Business Management*, EAI/Springer Innovations in Communication and Computing, https://doi.org/10.1007/978-3-030-82763-2_14

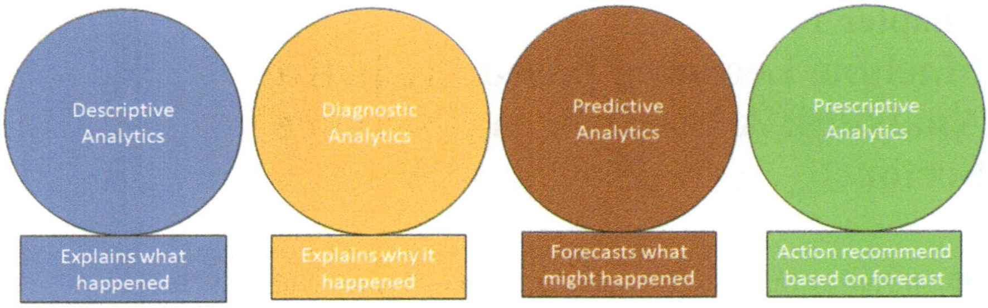

Fig. 14.1 Types of data analytics

14.1.2 Diagnostic Analytics

It is the type of analytics in which some input is taken from descriptive analytics and further some more statistical methodology is applied to find the reason for old events happening again [8]. This type of analytics is used to identified the reason for occurrence of past events.

14.1.3 Predictive Analytics

It is the type of analytics in which input is taken from descriptive analytics and diagnostic analytics and further some advanced level statistical methodology is applied to forecast the events that might happen in the coming future. Based on past and historical database the organization can predict the future [9–11] and it will help to take necessary action if future happening is not predicted as per business goal and vision.

14.1.4 Prescriptive Analytics

It is additional type of analytics to predictive analytics in which some suggestions and actions are proposed. On considering these suggestions the past failure can be mitigated and success can be achieved [12–14]. The main focus of this chapter is to explain the significance of predictive analytics in business informative decisions and how machine learning algorithms empower predictive analytics for focused decisions.

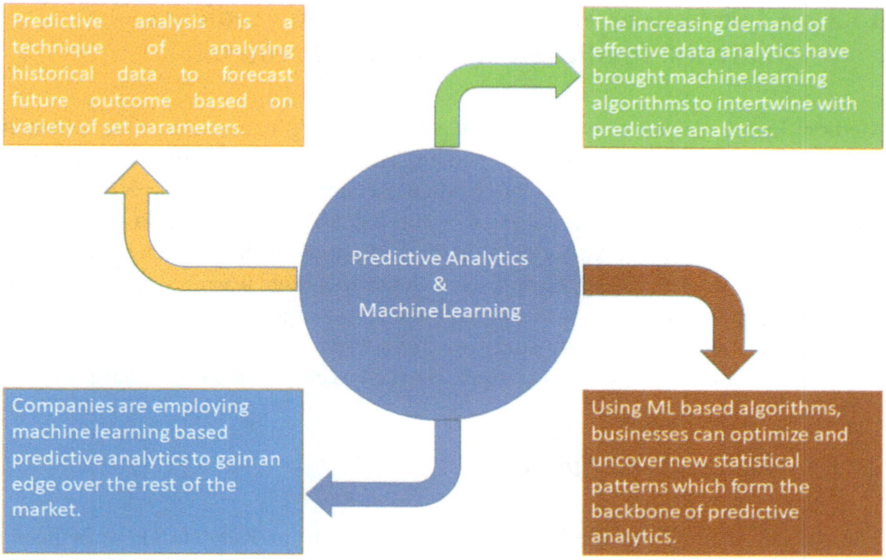

Fig. 14.2 Machine learning and predictive analytics intervention

14.2 Machine Learning Empowering Predictive Analytics

Predictive analytics is used for predictive future trends based on past and current data. Various statistical and modelling techniques are used to find hidden patterns and identify the trends to help in taking better informative decisions [15–16]. Earlier machine learning and predictive analytics are considered as two separate unrelated concepts.

The increasing demand for effective and précised decision for business brought together machine learning and predictive analytics. Machine learning algorithms are empowering predictive analytics to predict future decisions in more effective manner. In Fig. 14.2, we have shown how these two concepts are contributing to each other. There are many ML based algorithms like decision tree, random forest, linear regression, etc. used together with predictive analytical model to improve accuracy of prediction and provide good results to business owners.

The accuracy of prediction model can be enhanced in many dataset by combining more than one ML based algorithm with predictive analytical model.

14.3 A Use Case of Machine Learning and Predictive Analytics: Prediction of Insurance Premium

Medical insurance premium is expensive but there are many factors on which it depends and it varies from person to person. There are some important factors like age, gender, smoking status, no. of children, occupation, etc. Insurance company considers the values of these parameters and calculates the premium using some statistical and data analytical formula. In this case study we have taken an insurance company dataset from Kaggel and used some machine learning algorithm and completed predictive analytics to calculate the rate/charges of insurance premium.

For implementing this case study we have used linear regression model and sklearn python library to predict charges. So the target y-variable is 'Charges'.

14.3.1 Dataset Description

The dataset used for predictive analytics is taken from Kaggel. There is insurance.csv file that contains approximately 2000 records of various persons. It contains seven columns named as age, sex, bmi, children, smoker, region and charges (Table 14.1).

Considering these features step by step analytics is done for accurate prediction of insurance premium cost. As we mentioned above that our y-variable is charges. The min value is only ~1200 and max at ~63,000 with std. deviation ~12,000. This data deviates with high value so it could affect the model accuracy. The summary description of data is given below (Table 14.2).

Table 14.1 Dataset description with top five records

Age	Sex	BMI	Children	Smoker	Region	Charges
18	Male	34.1	0	No	Southeast	1137.011
34	Female	31.92	1	Yes	Northeast	37701.88
37	Male	28.025	2	No	Northwest	6203.902
59	Female	27.72	3	No	Southeast	14001.13
63	Female	23.085	0	No	Northeast	14451.84

Table 14.2 Brief summary of dataset

	#Age	#BMI	#Children	#Charges
Count no.	1400	1400	1400	1400
Mean	39.207025	30.663	1.094918	13270.422265
Std	15.012	6.0981	1.205493	12110.011237
Min	19.700	16.780	0	1321.450
25%	28	26.296	0	39750.140
50%	40	29.760	1.0000	8492.0330
75%	55	35.678	2.5000	179.814
Max	65	54.643	5.000	72670.282

14.3.2 Exploratory Data Analysis

Firstly with exploratory data analysis we will try to understand our dataset variable and how these are correlated with each other. The following pairplot chart is drawn from the given dataset (Fig. 14.3).

The pairplot helps use to identify the correlation between multiple features of the dataset. It is matrix of different kinds of charts where diagonal charts are bar plot charts. This matrix of chart is used to find the correlation between various features of dataset.

Similarly from heat map given in Fig. 14.4 signify that there is strong correlation between smoker status and average correlation with bmi and children.

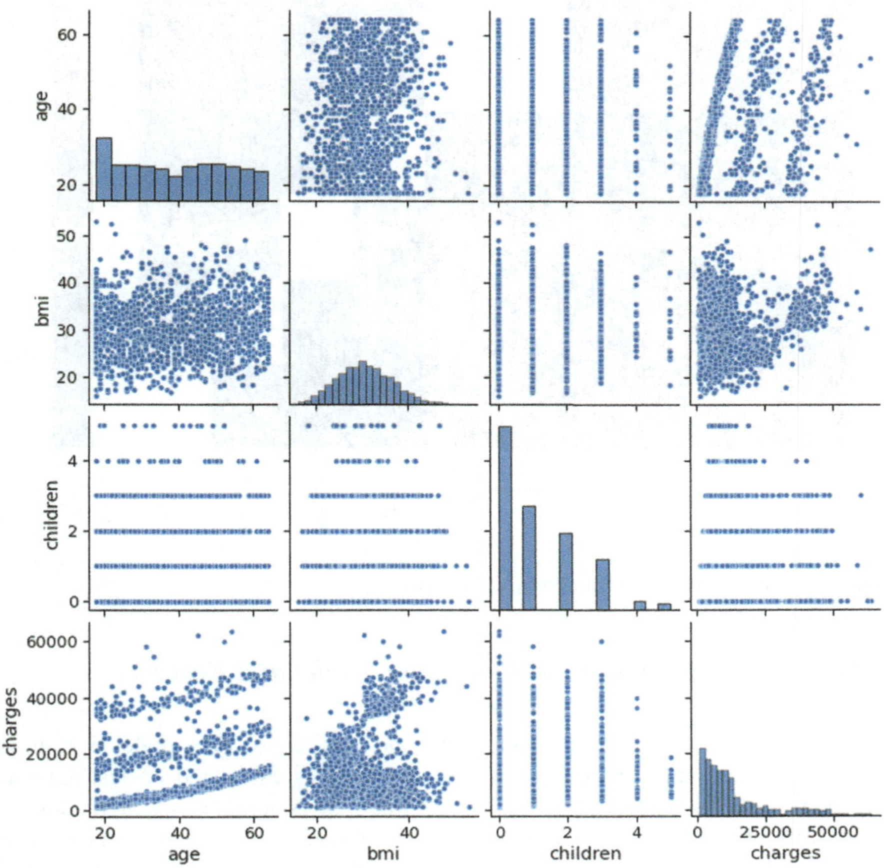

Fig. 14.3 Pairplot chart between main features of dataset

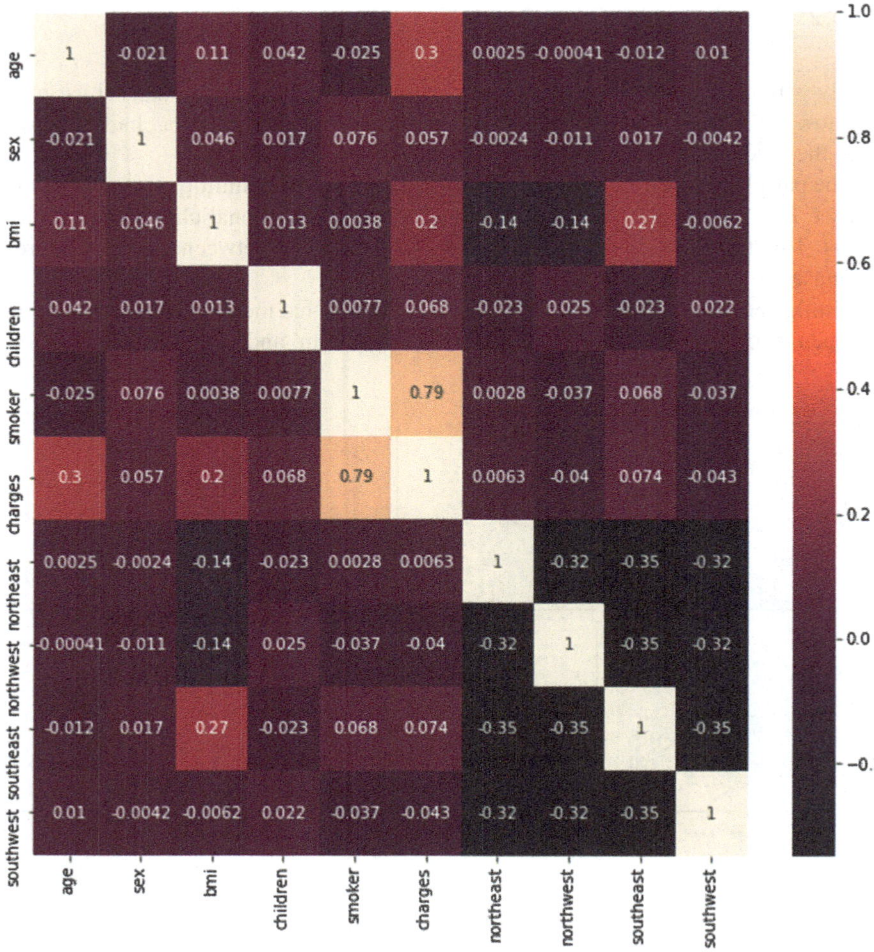

Fig. 14.4 Heat map for correlation between features

14.3.3 Implementation of Prediction Model and Results

The train and test data is split into 70% and 30% ratio. The model is trained on 70% dataset and accuracy parameters and error is calculated on test dataset. Random_statet is set at 42 which is arbitrary. So sample is taken randomly. MinMaxScaler is used to scale the data. Linear regression is used from sklearn library and scaled data is fitted to model. Model performance is evaluated using MAE, RMSe and MSE. Charges mean is ~13,000 and std. deviation ~12,000 is calculated. It shows the linear relationship has lots of noise and outliers and extreme values in dataset.

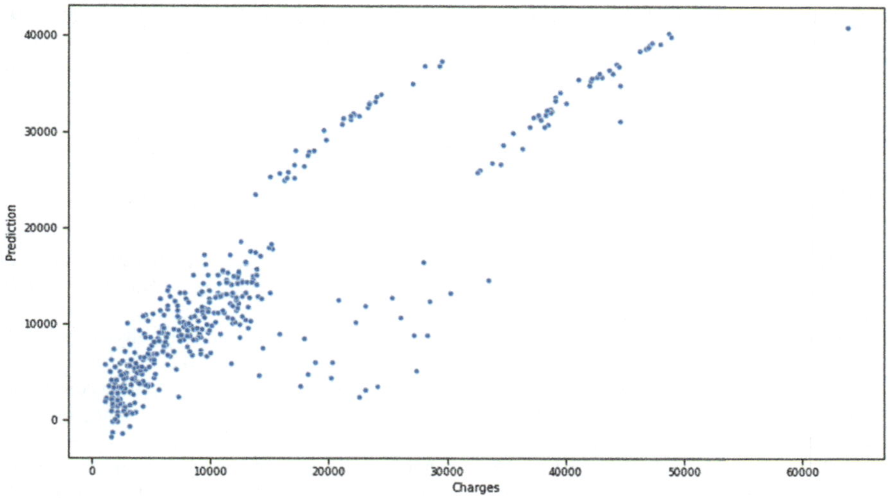

Fig. 14.5 Scatterplot for model accuracy

Table 14.3 Coefficient table of various parameters

Variable feature	Coefficient
Smoker	23628.367222
Children	424.119128
Bmi	348.906915
Age	261.296924
Sex	104.811823
region_northwest	−486.934610
region_southwest	−926.322908
region_southeast	−970.968839

In Fig. 14.5, the data is visualized on scatterplot. It is clearly shown some outliers and extreme values present in dataset produce error in linear regression. It shows the similarity and difference between actual and predicted output (Table 14.3).

From the above coefficient table, the most contributed factor in charges is smoker. Increase in 1 unit of smoker the charges will increase approximately to ~$23,600. Other most important factors are BMI and age. As we were expecting that smoking plays the most correlated feature to charge.

14.4 Conclusion

The four major types of analytics have important roles deep down in business data warehouse and take out important trends and patterns to analyse the whole business statistics. Most of the businesses use predictive analytics majorly to take informative decision future growth. We have described in this chapter that how machine learning is given excellent support to predictive analytics to get desired outcome from historical dataset.

References

1. P.K.D. Pramanik, S. Pal, M. Mukhopadhyay, Healthcare big data: A comprehensive overview, in *Intelligent Systems for Healthcare Management and Delivery*, ed. by N. Bouchemal, (IGI Global, Hershey, PA, 2019)
2. A. Shastri, R. Biswas, A framework for automated database tuning using dynamic SGA parameters and basic operating system utilities. Database Syst. J. **3**, 4 (2012)
3. K. Kshitiz, Shailendra, NLP and machine learning techniques for detecting insulting comments on social networking platforms. International Conference on Advances in Computing and Communication Engineering (ICACCE), IEEE, 2018, pp. 265–272
4. I. Khanchi, E. Ahmed, H.K. Sharma, Automated framework for real-time sentiment analysis, in International Conference on Next Generation Computing Technologies (NGCT-2019)
5. I. Khanchi, N. Agarwal, P. Seth, P. Ahlawat, Real time activity logger: A user activity detection system. Int. J. Eng. Adv. Technol. **9**(1), 1991–1994 (2019)
6. A. Bhushan, V. Jain, T. Singh, K. Munjal, Model for automated database tuning framework. Int. J. Contr. Theory Appl **9**(22), 347–357 (2016)
7. K. Kshitiz, H. Singh, P. Kukreja, Detecting hate speech and insults on social commentary using nlp and machine learning. Int. J. Eng. Technol. Sci. Res. **4**(12), 279–285 (2017)
8. P. Ahlawat, S.S. Biswas, Sensors based smart healthcare framework using internet of things (IoT). Int. J. Sci. Technol. Res. **9**(2), 1228–1234 (2020)
9. B.A. Thakkar, M.I. Hasan, M.A. Desai, Health care decision support system for swine flu prediction using näive bayes classifier, in 2010 International Conference on Advances in Recent Technologies in Communication and Computing, Kottayam, 2010, pp. 101–105
10. R. Vaishya, M. Javaid, I.H. Khan, A. Haleem, Artificial Intelligence (AI) applications for corona virus disease pandemic. Diabetes Metab. Syndr. Clin. Res. Rev. **14**(4), 337–339 (2020)
11. G. Dangi, T. Choudhury, P. Kumar. A smart approach to diagnose heart disease through machine learning and springleaf marketing response, in 2016 International Conference on Recent Advances and Innovations in Engineering (ICRAIE), 2016, pp. 1–6
12. P. Kumar, T. Choudhury, S. Rawat, S. Jayaraman. Analysis of various machine learning algorithms for enhanced opinion mining using twitter data streams, in 2016 International Conference on Micro-Electronics and Telecommunication Engineering (ICMETE), 2016, pp. 265–270
13. T. Choudhury, G. Dangi, T.P. Singh, A. Chauhan, A. Aggarwal, An efficient way to detect credit card fraud using machine learning methodologies, in Proceedings of the 2nd international conference on green computing and internet of things, ICGCIoT, 2018
14. R. Kataria, T. Sharma, T. Choudhury, P. Kumar, Image enhancement of old manuscripts using machine learning, in Proceedings of the 2017 3rd International Conference on Applied and Theoretical Computing and Communication Technology, ICATccT 2017, 2018

Chapter 15
Machine Learning Applications in Decision Intelligence Analytics

Sanika Singh, Aman Anand, Saurabh Mukherjee, and Tanupriya Choudhury

15.1 Introduction

There are various parts in our life where we are surrounded by machine learning algorithms from our mobile to laptops or somewhat from automated cooler to automated washing machines. Hence, along with these small things, machine learning has huge applications in individuals' lives.

Traditional Programming: Data and program are run on the computer system to generate the output. Machine Learning: Data and output are run on the computer to create a program. This program can be used in traditional programming. Machine learning let the computers code themselves. If programming is automation or programmed in such a way that there is less human interference, then machine learning [1–5] algorithms are used to automate the process of automation. You can correlate machine learning with farming or gardening. Seeds are algorithms, nutrients are data, the gardener is you, and plants are the programs (Fig. 15.1).

Machine learning is popularly known as the study of algorithms developed by programmers which improves automatically through experiences or which is popularly known as the part of training. There are many different types of learning

S. Singh · S. Mukherjee
Department of Computer Science and Engineering, Banasthali Vidyapith, Vanasthali, India

A. Anand
Department of Computer Science and Engineering, Sharda University, Noida, India

T. Choudhury (✉)
School of CSE, University of Petroleum and Energy Studies, Dehradun, Uttarakhand, India
e-mail: tanupriya1986@gmail.com

P. M. Jeyanthi et al. (eds.), *Decision Intelligence Analytics and the Implementation of Strategic Business Management*, EAI/Springer Innovations in Communication and Computing, https://doi.org/10.1007/978-3-030-82763-2_15

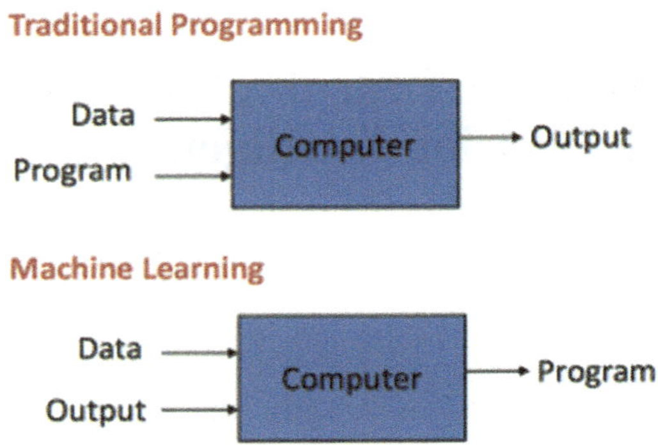

Fig. 15.1 Pictorial difference between traditional programming and machine learning

that a machine learning model used to perform, among these types the most popular ones are learning problems, (Fig. 15.2) like:

1. Supervised learning
2. Unsupervised learning
3. Semi-supervised learning
4. Reinforcement learning

Machine learning is used to solve problems in six different steps, which are mentioned as follows:

- Feature extraction + domain knowledge
- Feature selection
- Choice of algorithms
- Training
- Choice of metrics/evaluation criteria
- Testing

Recently, machine learning has earned a lot of popularity and is discovering its path through major domains such as medical, banking system, entertainment like Netflix, Amazon, etc. There is so much we can do with it, see "How Google Uses Machine Learning and Neural Networks to Optimize Data Centers."

Decision intelligence (DI), as they claim, resolves the world's most difficult issues. It brings human decision-makers together to technology such as artificial learning, IA, deep learning, simulation of visual decisions, complex device modeling, big data, UX architecture, mathematical analysis, enterprise intelligence, business process management, causal inference, proof-based analysis, and more.

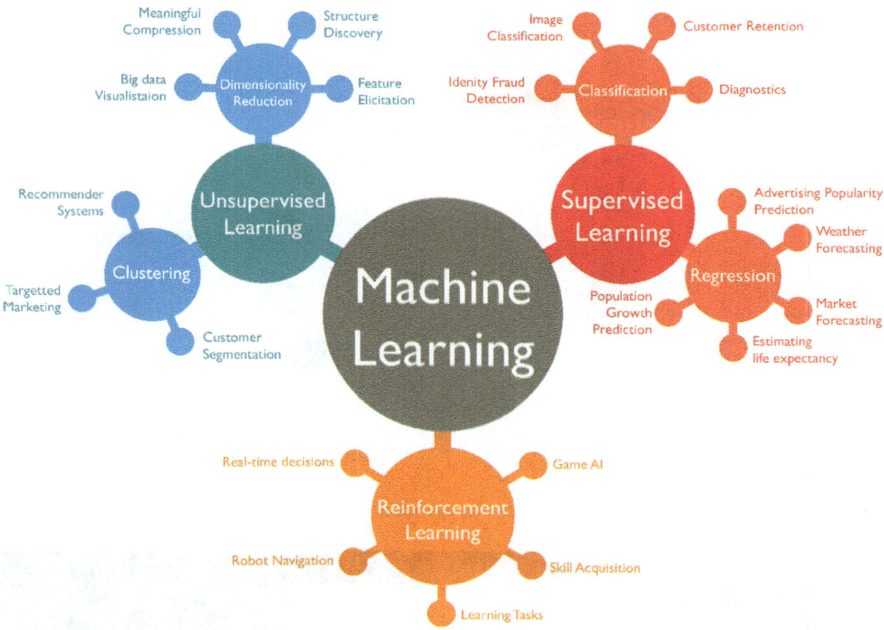

Fig. 15.2 Uses of different types of machine learning parts

15.1.1 Application of Machine Learning

15.1.1.1 Virtual Personal Assistants (VPA's)

Google assistants, Cortana, Alexa, Siri are some of the examples of VPA, i.e. virtual personal assistants. In our day-to-day life we are surrounded with such things and we all are using these at some point of our life. Now, these help improve our life in a great number of ways [6]. Like, for example: You can tell them to call someone, you can tell them to play some music, you can tell them to even schedule an appointment. So, how do these things actually work, below are the which they follow:

1. They record what we are saying.
2. Send it to a server which is usually in the cloud.
3. Decode it with the help of machine learning and neural networks.
4. And, then provide us with the required output (Fig. 15.3).

So, if you have been ever noticing that these systems are not working properly without the internet, it is because of the server it will not be able to contact it. And, according to recent research about 38% of the consumers are currently using virtual personal assistants in their devices (Fig. 15.4).

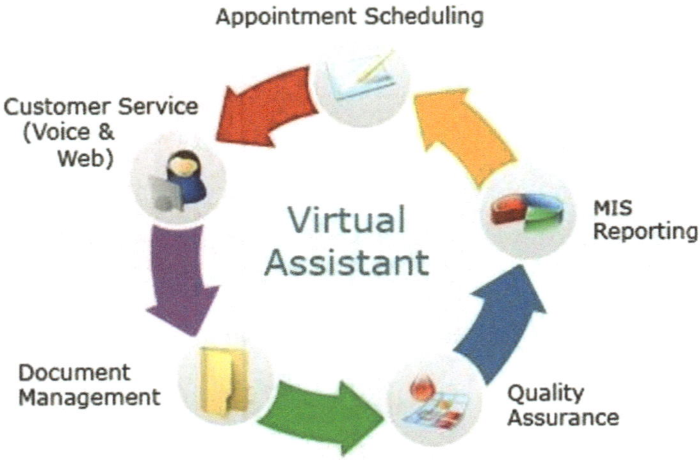

Fig. 15.3 Tasks performed by virtual personal assistants

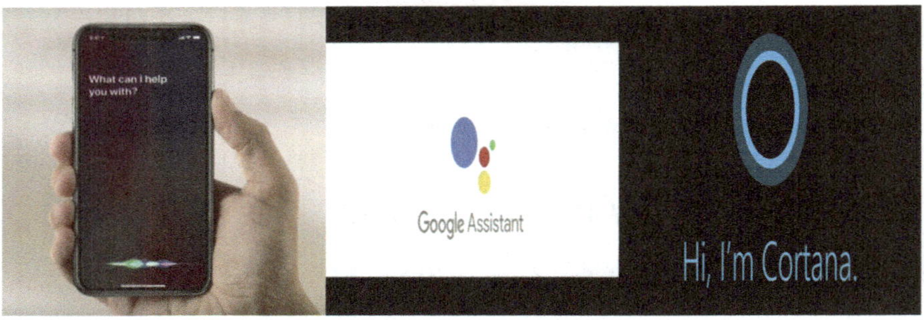

Fig. 15.4 Examples of virtual personal assistants

Advantages

- Open applications.
- Trace nearby locations.
- VPAs are used to book movie tickets.
- Find out quick information about any news, weather conditions, financial conditions of any country, any pandemic.

Disadvantages

- It will not work without the internet.
- Hindi language is not fully supported by VPAs.
- Hang on your mobile phones for longer.
- Maximum usage of batteries.
- Larger amount of heat generation by cellphones.
- High usage of data.

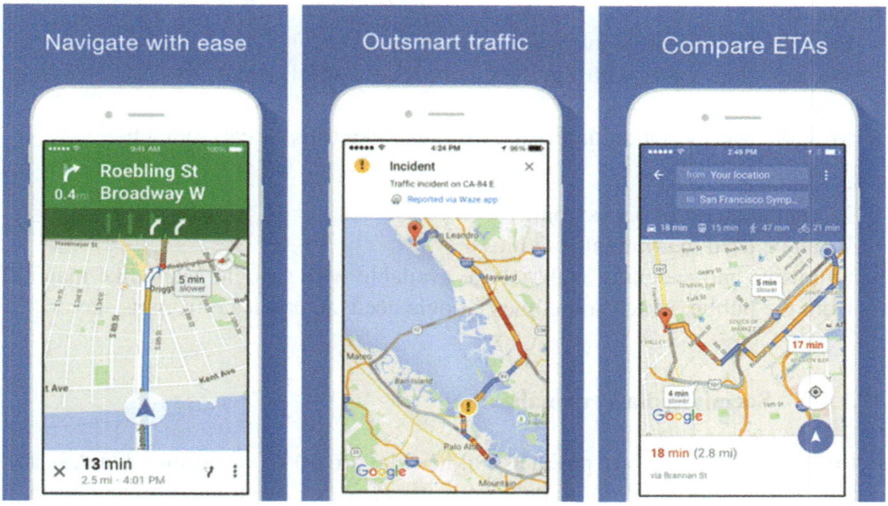

Fig. 15.5 Google maps cases: Blue line: predicts for less amount of traffic, Yellow line: predicts to move slowly, Red line: predicts for heavily congested traffics

15.1.1.2 Traffic Predictions

Now, suppose we want to travel from one place to another in a location where we are totally new in this region. What we probably do? We will definitely use Google Maps to travel from one location to another in that region (Fig. 15.5).

Google Map is a web mapping (the process of accessing the maps provided by the geographic information system (GIS) in World Wide Web (WWW)) services provided to us by Google. It provides us with satellite images, aerial photography, street maps, 360° interactive extensive view of roads, real-time traffic monitoring, and route or direction planning for traveling by foot, cars, bicycles and air, or by using public transportations. So, how actually Google maps predicts which direction is best among the huge number of options with Google [7].

Google Maps predicts whether the traffic is clear, slow moving, or heavily congested with the help of machine learning algorithms and two important factors:

- Average time taken on specific days at specific times on that route.
- Real-time location data of vehicles from Google Maps application and sensors.

Some of the popular map services available are as follows:

- Bing Maps
- Mpas.Me
- Here we go

Advantages

- Bounty of Information available.
- We can share maps with families, colleagues, which makes meetings easier to organize.
- Multiple transportation modes are available like trips by car, foot, etc.

Disadvantages

- Limited amount of perfection.
- It can be used in criminal activities.
- Offensive and shocking material are available with Google Maps like material from the public, which sometimes Google needs to eradicate from its server.

15.1.1.3 Social Media Personalization

Let us say we want to buy products online (like drone) or something. On Amazon or any other online platform to buy products we search about the products (drone), and it costs about 1 lakh. So, at that point of time one is not interested in buying it.

Now, let us say one visits Facebook and they found the same products (drone) advertisement. Next time one visits YouTube there they see advertisements and when they visit Instagram again the advertisements are there also.

So, here with the help of machine learning algorithms of content filtering, collaborative filtering, and hybrid filtering of product recommendation techniques our search engine will understand that the customer is keen interested in a particular product and hence, the search engine targeting us with these advertisements with the help of machine learnings [6] (Fig. 15.6).

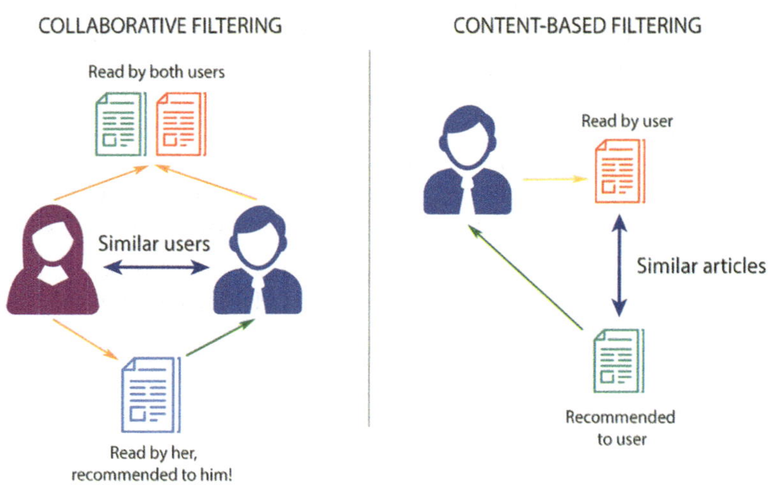

Fig. 15.6 Product recommender system techniques

Advantages

- Social media marketing helps the products to increase their brand awareness.
- Because of its good services it provides better customer satisfaction.
- As registering in any e-commerce website is almost free of cost and by using SMM experts it is cost-effective.
- Provide high-valued content for the products and by the use of social media without lots of signing in into the website it is useful in providing the increased inbound traffic.
- By using SMM tools we can collect customer data which is useful in gaining market insight.

Disadvantages

- Social media marketing tools help to gain knowledge about the competitors.
- To need a huge amount of traffic control and to enrich business to a higher level, one requires qualified employees.
- Since it is a long-term investment so one cannot desire to return instantly.
- Here, customers get a chance to provide review for products, so it may tarnish brand value.
- It may be as customers may pass your pages and look for another one, so it is a time-consuming process.

Applications

- Personalized video/picture marketing techniques.
- Personalized retargeting advertisements through social media platforms.
- Directly messaging on Twitter handle.
- Personalized review asking on social media platforms.
- Real-life operations, based on data achieved socially.

15.1.1.4 Email Spam Filtering

As everyone knows that there is "SPAM" section in every email services. So, how does the email know that a particular message is SPAM or not? So, Gmail has an entire collection of emails which have already been labeled as SPAM or not SPAM. By analyzing this data Gmail is able to find out some characteristics which are able to determine SPAM messages, like the word "lottery," "Winner," etc. From then on in a new email the constituent box goes through a few emails SPAM filters to decide whether it is SPAM or not. A filtering solution is put to email systems which utilize a set of protocols (rules) to regulate which incoming mails are spam and not spam. The methodology followed in the process is shown in Fig. 15.7.

Some of the most popular SPAM filters used by Gmail are:

- Content Filters—Analyze the contents within mail to regulate whether it is spam or not.
- Header Filters—Analyze the email headers in search of fake information.
- General Blacklist Filters—It terminates all electronic mails that come from a blacklisted (blocked) portfolio of known spammers.

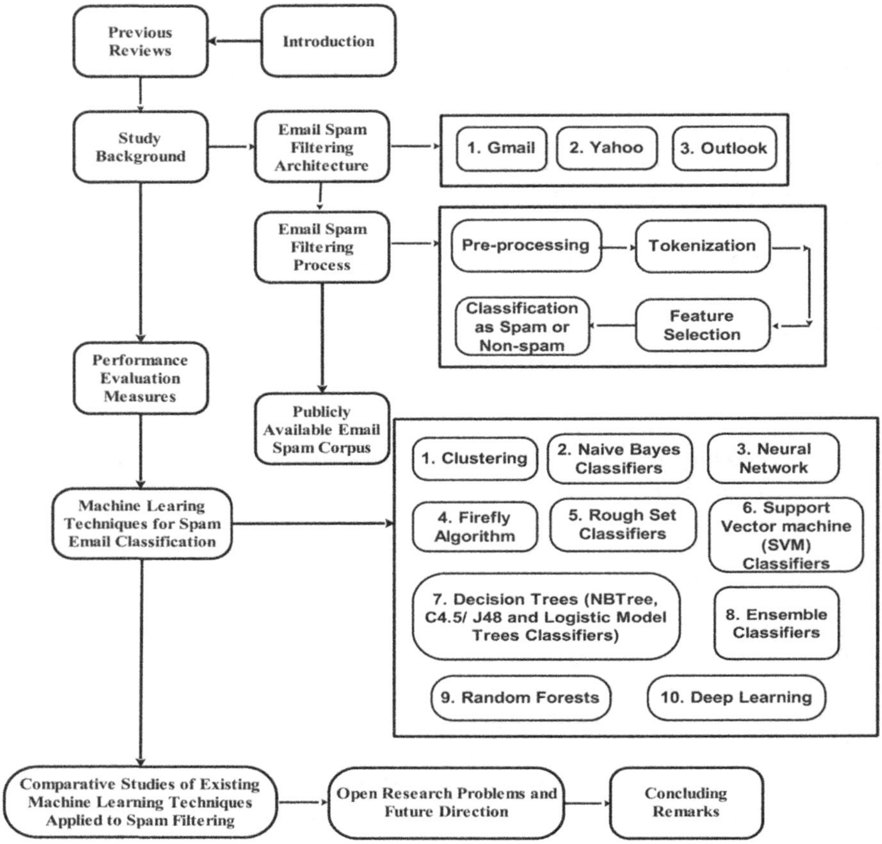

Fig. 15.7 Process of working of email spam filtering using machine learning algorithms

- Rules-based Filters—It uses user-defined scenarios—like a particular sender or particular words in the subject or body of emails, so that it will block spam.
- Permission Filters—It needs anyone to send a mail to be pre-approved by the receiver.
- Challenge-Response Filters—It needs someone sending a mail to put code (OTP) in order to get permitted to send email.

Advantages

- Spam emails are not just righteous advertising elements which may be carriers of hazardous computer viruses or worms. Just by clicking it once on the inappropriate messages can debilitate your web. Filters can be used as one of the suitable firewalls.
- In addition to hazardous viruses, hackers or crackers can also get access to the system through beneficial looking mails. A filter which is used to block spam mails from outreaching inbox can save chief data for a user.

- Anti-spam software and applications can be customized according to user needs. One can create a blacklist of mail addresses that usually send spam messages.
- Spam filtering can be used to save time. The time saved can be used to increase productivity of the user.

Disadvantages

- Spam—Commercial email or "spam" irritates consumers. The "click through rate" for non-targeted mails is more likely to be very less.
- Undelivered Emails—Poorly designed mails may not get transmitted. Mails that use some spam keywords or characters in the subject heading or content of the email, e.g. £££s, FREE, click here, are likely to be filtered out by mail software and ISPs.
- Design problems—Email must be planned so that its appearance is as it should across multiple gadgets and email providers.
- Size Issues—Files need to be very small to get downloaded quickly. Mails which contain many pictures may take time to load, frustrating the audience and they lose their interests.
- Resources and skills—For a successful email promotion we must ensure that we have the copyright, design, and marketing list.

Analysis of Different Machine Learning Algorithms

- Naive Bayes Algorithm is mostly used in email spam filtering technique as it is one of the popular statistical tools for email spam filtering (Fig. 15.8). Naive Bayes spam filtering algorithm is a standard method for conducting spam messages which can customize itself to the email requirement of individual clients and provide very less erroneous positive spam detection outlay which are mostly acceptable to the clients.

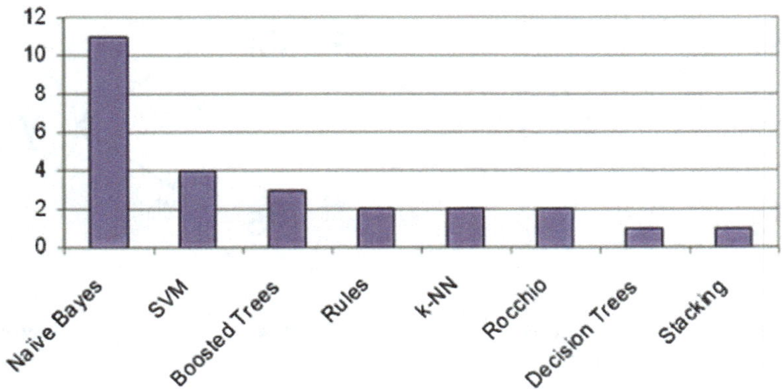

Fig. 15.8 Algorithms used in email spam filtering

15.1.1.5 Online Fraud Detection

There are several ways by which online fraud detection takes place [8, 9]. For example:

1. Identity Theft—It is the case where they steal your identity.
2. Fake Accounts—It only lasts for how long the transaction takes place and it stops existing after that.
3. Man in the Middle Attacks—It is the case where they steal your money while the transaction is taking place.

   - Feed-Forward Neural Network helps to determine whether the transaction is genuine transaction or raud transaction. So, what happens with a feed forward neural network is that the output is converted into hash values and these values become the input for the next round. So, for every real transaction that takes place forms a specific pattern. A fraud transaction stands out because of the significant changes that it would cause with the hash values (Fig. 15.9) (Table 15.1).

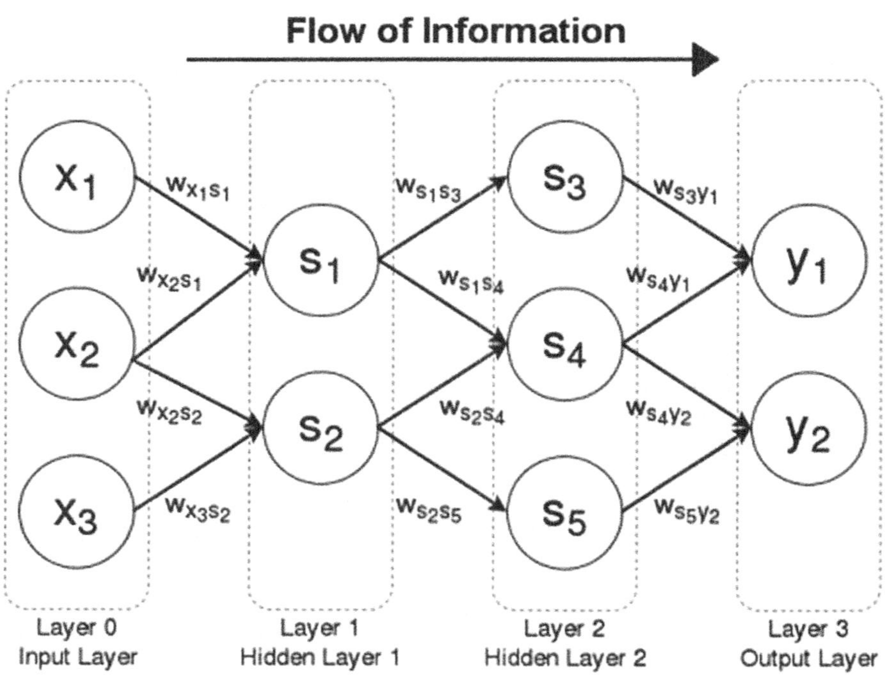

Fig. 15.9 Flow of information by using hidden layer in feed forward neural network

Table 15.1 Comparison of advantages and disadvantages of online fraud detection techniques (shown in Fig. 15.9)

S. no.	Fraud detection algorithms	Advantages	Disadvantages
1.	K-nearest neighbor algorithm (KNN)	It is used to determine anomalies in the target instance and is easy to implement [10]	It is suitable to detect frauds with the limitations of the memory
2.	Hidden Markov model (HMM)	It can detect the fraudulent activity at the time of the transaction [11]	It cannot detect fraud with a few transactions
3.	Neural network (NN) (Fig. 15.10)	It is used to learn the previous behavior and is used to detect the real-time credit card frauds	It has many sub-techniques. So, if they pick up this it is not suitable for credit card fraud detection, the performance of the method will decline
4.	Decision tree (DT)	It can handle nonlinear credit card transactions as well	DT cannot be able to detect real-time fraud detection
5.	Deep learning (DL)	Analysis of unsupervised data done by this method	Library of DL does not cover all algorithms. As it is mostly used in image recognition and all

- Feed Forward Neural Network: It is an artificial neural network which forms connections between the junction that do not form a cyclic pattern. As such, it is distinct from its successor: recurrent neural networks. It was the initial and simpler kind of artificial neural networ devised (Fig. 15.10).

15.1.1.6 Assistive Medical Technology

Medical technology has innovated with the use of machine learning to diagnose diseases. We are able to analyze or predict 2D CT (computed tomography) scans and come up with 3D models that help to analyze where there are lesions in the Brain. It works equally well for brain tumor and ischemic stroke lesions. This can be further used as fetal imaging and cardiac analysis too. The methodology followed in the process is shown in Fig. 15.11.

Medical "fields" where Machine Learning can be "applied"

- Disease identification
- Personalized treatment
- Drug discovery
- Clinical research
- Radiology

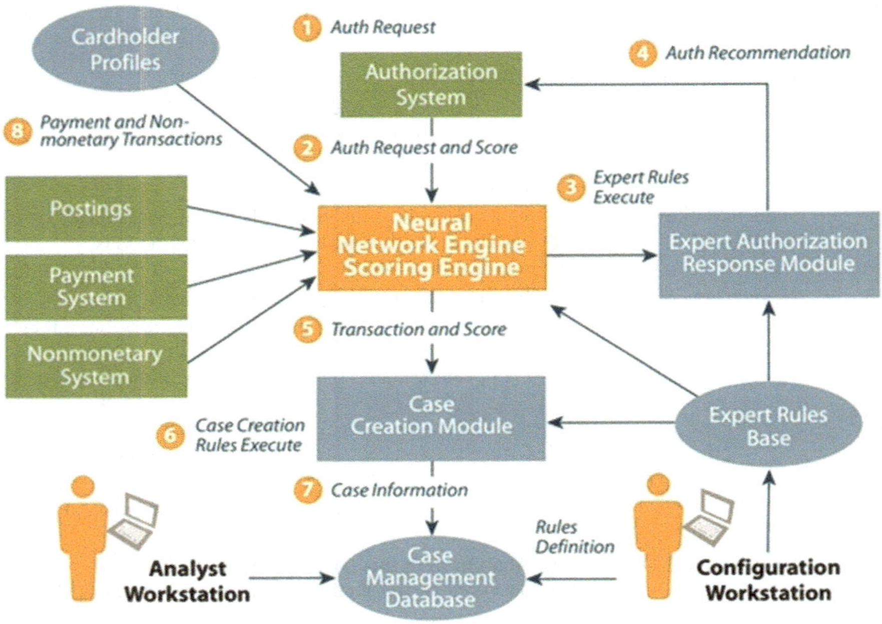

Fig. 15.10 Working of machine learning algorithm (neural network) in fraud detection

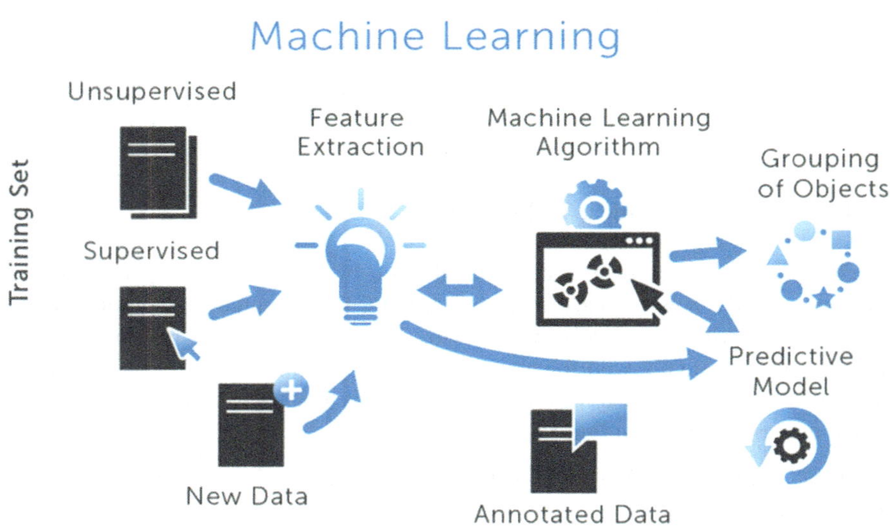

Fig. 15.11 Machine learning model used for identification of disease

Advantages of Assistive Medical Technology Are

1. In the research field it helps in discovering drugs as well as clinical trials of the drugs.
2. It helps also in the customized medical treatment.
3. It also helps in retrieval of the smart health management of the old documents.
4. Identification of lesions on a broader classification.
5. Robotic surgeons help in assistance of the treatment of the diseases.

Disadvantages of Assistive Medical Technology Are

1. Need larger data sets to train a model on higher accuracy.
2. Rate of error generation is higher.
3. Learner personnel are required to operate in medical.
4. Physical conditions need to meet up with the model.
5. Trained to work for a longer duration without getting tired.

15.1.1.7 Automatic Translation

Say, you are in Spain with your smartphone. And there is something written on the wall or anywhere else and you do not know then here comes up the automatic translation which helps one to get to learn about different scripts available (Fig. 15.12).

The technology behind this is sequence-to-sequence learning which is similar to the algorithms used in chatbots. Here, image recognition is used with the help of CNN (convolutional neural network) and the text is identified by using OCR (optical character recognition) [12, 13]. Further, sequence-to-sequence algorithm is used to translate the text from one language to another. The use of software application

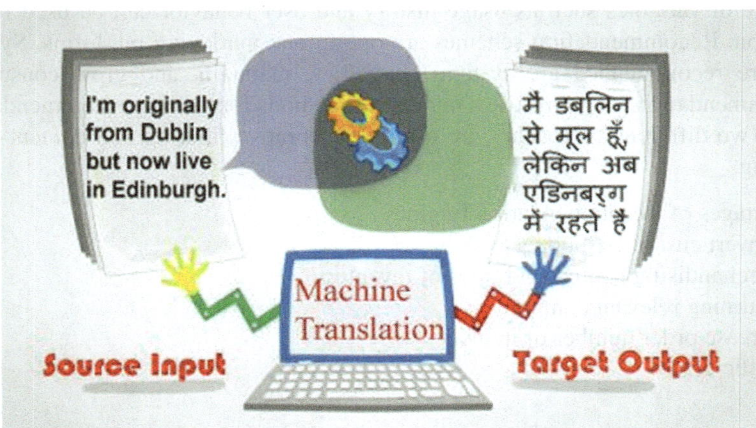

Fig. 15.12 Applications of machine translation as well as its some limitations

programs which nowadays specifically developed to translate both spoken and written texts from one language to another.

Advantages of Automatic Translation

- When time is an important factor, automatic translation can save time. You do not need to spend time on searching over dictionaries to get the words translated. Instead, the application can translate the content fastly and make available high-quality results to the applicant in zero time at all.
- It is comparatively cheaper.
- Confidentiality is another factor which makes machine automatic translation more acceptable. Giving most sensitive data to a translator might be a risk, while with automatic translation our information is protected.
- Translators usually convert text which is in any language so that there is no such major concern while an experienced translator specialized in one particular field only.

Disadvantages of Automatic Translations Are

- Accuracy not granted.
- It will not be able to concentrate on a context and solve ambiguity and neither makes use of mental outlook like a human translator can.

15.1.1.8 Recommendation Engines

A recommendation engine is a data filtering application with machine learning algorithms that suggest to a certain consumer or client the most pertinent things. It is based on the concept of discovering patterns in user data that can be indirectly or directly obtained. An engine of recommendation is a system which offers goods, services, and user information based on data analysis [14]. Nevertheless, a range of variables such as usage history and user behavior can be used for the decision. Recommendation schemes are one-to-one marketing platforms. Systems that are recommended are systems that draw, maintain, and grow consumers. Recommendation programs use a number of methods for making recommendations [15]. Two different approaches are used: collaborative filtering and content-based filtering.

Advantages of Recommendation Engines

- Convert customer shoppers
- Merchandising control and rules of inventory
- Returning relevant content
- Increase order number of items
- Traffic drive

Disadvantages of Recommendation Engines

- Items unpredictable
- User preferences change
- Data modification
- Data missing

15.2 Summary and Conclusion

Machine learning technique is a swiftly developing phase in computer science and information technology. It is applicable in nearly each and every alternative field of learning and is hitherto being applied commercially as ML can sort out problem statements which are too difficult or extremely time consuming for humans to work out.

This chapter has established us in machine learning. Now, we know that ML is a technology of making machines to learn and to perform to carry out ventures a human brain can drain, which is faster and superior than a usually human-being. Today as there is advancement in technologies the machines can beat human giant brains in games like: Chess, AlphaGo, which are contemplated to be very tough. Machines which can be learned to take action that humans usually take in specific areas of interest can aid humans to live better lives.

Machine learning can be grouped into supervised or unsupervised learning. If minimum quantities of data and information are there and distinctly labeled up data for learning, then choose the supervised learning model. Unsupervised learning models are generally accessed for high-quality performance and consequences for big data sets. If a big data set is easily accessible, opt for deep learning (DL) technology. Applications of neural embedded networks are their limitations which are persisting.

Firstly we know that, when it comes to development of ML modules the alternative of different development languages, IDEs and Platforms are needed. Other things that are required to perform are to start learning and training each ML technology. The topic is very vast, it means that there is breadth, but if carried out in depth, each and every topic can be in the human brain within a few times. Every topic is free of each other. Take into observation only one topic at a time, learn it, practice it, and apply the algorithm/s in it by using a choice of your language. It is the one of the best ways to start implicating ML. Practicing a topic at a time, easily one would access the width, finally needed by an ML expert.

References

1. P. Kumar, T. Choudhury, S. Rawat, S. Jayaraman, Analysis of various machine learning algorithms for enhanced opinion mining using twitter data streams, in 2016 International Conference on Micro-Electronics and Telecommunication Engineering (ICMETE), 2016, pp. 265–270
2. S. Jayaraman, T. Choudhury, P. Kumar, Analysis of classification models based on cuisine prediction using machine learning, in 2017 International Conference on Smart Technologies for Smart Nation (SmartTechCon), 2017, pp. 1485–1490
3. A. Verma, A. Shukla, T. Choudhury, A. Chauhan, Recommendation system for prediction of tumour in cells using machine learning approach, in International Conference on Recent Developments in Science, Engineering and Technology, 2019, pp. 212–222
4. U.V. Singh, E. Gupta, T. Choudhury, Detection of rheumatoid arthritis using machine learning, in 2019 International Conference on Computational Intelligence and Knowledge Economy (ICCIKE), 2019, pp. 25–29
5. V. Kumar, T. Choudhury, A.S. Sabitha, S. Mishra, Summarization using corpus training and machine learning, in *Emerging Trends in Expert Applications and Security*, (Springer, Singapore, 2019), pp. 555–563
6. S. Angra, S. Ahuja, Machine Learning and its application, in International Conference on Big Data Analytics and Computational Intelligence (ICBDAC), 2017
7. V.L. Hans, V.H. Chris, Short term traffic and travel time prediction models, artificial intelligence applications to critical transportation issues. Transport. Res. **168**, 22–41 (2012)
8. P. Sahil, G. Dishant, A. Mehak, K. Ishita, J. Nishtha, Comparison and analysis of spam detection Algorithms. Int. J. Appl. Innov. Eng. Manage. **2**(4), 1–7 (2013)
9. S.M. Lee, D.S. Kim, J.H. Kim, J.S. Park, Spam detection using feature selection and parameters optimization, in 2010 International Conference on Complex, Intelligent and Software Intensive Systems, 2010, pp. 883–888
10. S. Bozkurt, G. Elibol, S. Gunal, U. Yayan, A comparative study on machine learning algorithms for indoor positioning, in International Symposium on Innovations in Intelligent SysTems and Applications, 2015
11. R.R.R. Robinson, C. Thomas, Ranking of machine learning algorithms based on the performance in classifying DDoS attacks, in IEEE Recent Advances in Intelligent Computational Systems (RAICS), 2015
12. O. Abdel-hamid, L. Deng, D. Yu, Exploring convolutional neural network structures and optimization techniques for speech recognition, in Proceedings of the Annual Conference of the International Speech Communication Association, INTERSPEECH, 2013, pp. 3366–3370
13. M. Kumar, S.R. Jindal, M.K. Jindal, G.S. Lehal, Improved recognition results of medieval handwritten Gurmukhi manuscripts using boosting and bagging methodologies. Neural. Process. Lett. **50**, 43–56 (Sep. 2018)
14. S. Singh, A. Anand et al., A recommender system for component-based applications using machine learning techniques, in International Conference on Machine Intelligence and Data Science Applications, October 2020
15. J. Lu, D. Wu, M. Mao, W. Wang, G. Zhang, Recommender system application developments: a survey. Decis. Support Syst. **74**, 12–32 (2015)

Chapter 16
Demystifying Behavioral Biases of Traders Using Machine Learning

Hardeep Singh, Anurag Singh, and Era Nagpal

16.1 Introduction

Individuals' biases when studied in the field of behavioral finance, integrate psychology and finance. Behavioral finance is relatively new and has emerged in the mid-to-late 1980s. Behavioral finance [1] emerged as a branch of knowledge to answer the existing puzzles such as equity premium puzzle and dividend payout puzzle in corporate finance. Behavioral finance considers that individuals are not always rational and challenges the assumption of rationality. Amos and Tversky contributed significantly in the field of behavioral finance [2] and stated the presence of various behavioral biases. Behavioral finance helps to look at the problem with a pragmatic approach as it does not assume the parties involved in the decision-making to be rational. Although research in behavioral finance is still at its nascent stage, yet the research is gaining momentum. A diverse set of issues like the capital structure decisions of the firms, capital budgeting decisions, dividend policy decisions, mergers and acquisitions, and many more are studied in behavioral finance.

Behavioral biases in finance are categorized into two main parts: cognitive biases and emotional biases. The presence of biases results in sub-optimal decisions and deviates the firm from the value-maximizing principle. A lot of literature in behavioral finance is borrowed from the field of psychology and the application of this research is showing promising results. This chapter focuses on the cognitive biases of traders in the stock markets and highlights the contribution of techniques

H. Singh (✉) · A. Singh
DoGM, School of Business, University of Petroleum and Energy Studies (UPES), Dehradun, India

E. Nagpal
Chitkara Business School, Chitkara University, Punjab, India

P. M. Jeyanthi et al. (eds.), *Decision Intelligence Analytics and the Implementation of Strategic Business Management*, EAI/Springer Innovations in Communication and Computing, https://doi.org/10.1007/978-3-030-82763-2_16

of machine learning to deal with the cognitive biases of the traders. Traders are involved in a range of activities for the sale and purchase of financial securities. The presence of biases in traders' behavior deteriorates their performance and leads to underutilization of their skill. The experience, peer group, advertisement, rumors, and heuristics are the main sources of biases among traders.

Ramiah et al. [3] documented the presence of biases among traders and stated the existence of noise traders in the market. Researchers acknowledged the model to quantify and understand the noise trader risk. Several researchers have contributed to the growing literature in the field of behavioral corporate finance. We have applied the framework of behavioral biases developed in the literature to the behavior of traders and affirmed the application of specific techniques of machine learning to reduce the impact of behavioral biases on the decision-making of traders. The ultimate objective is to provide a recommendation for traders to increase their performance and for them to make sound decisions using machine learning tools. Thaler contributed significant work in the field of behavioral finance and some of the noteworthy works include [4, 5, 6]. Shefrin has also contributed in the field of behavioral corporate finance such with studies like [1, 7–10].

Figure 16.1 depicts cognitive biases among traders and the relevant machine learning techniques to help traders overcome those cognitive biases. The following section of the present chapter provides insights into the stated biases along with the use of machine learning to improve decision-making with the end objective of improving the performance of the traders.

Machine learning algorithms assist a computer in performing a variety of tasks. Also, machine learning is an integral part of modern artificial intelligence (AI). The behavior of agents (traders in the present chapter) is integrated with the data environment and the desired integration helps the agents to make sound decisions. Traders can use machine learning to mimic the world and learn from the past. As traders learn from experience and their enhanced knowledge results in heuristics and biases, the presence of such biases in the machine learning technique is negligible. The following section presents various cognitive biases among traders and the relevant techniques from machine learning to mitigate or eliminate such bias. Supervised, unsupervised, and reinforcement techniques of machine learning are presented in the following section. Supervised machine learning techniques focus on labeling pairs or providing a training set of features. The unsupervised machine learning techniques provide only the data set without any assistance for interpretation. Reinforcement learning techniques provide partial feedback and are classified as either optimization tasks or learning objects from the behavior.

16.1.1 Confirmation Bias

Confirmation bias deals with individuals' tendency to favor the information that agrees with their beliefs and ignore the contradictory information. [11, 12] documented the presence of confirmatory bias among traders. Traders attach undue

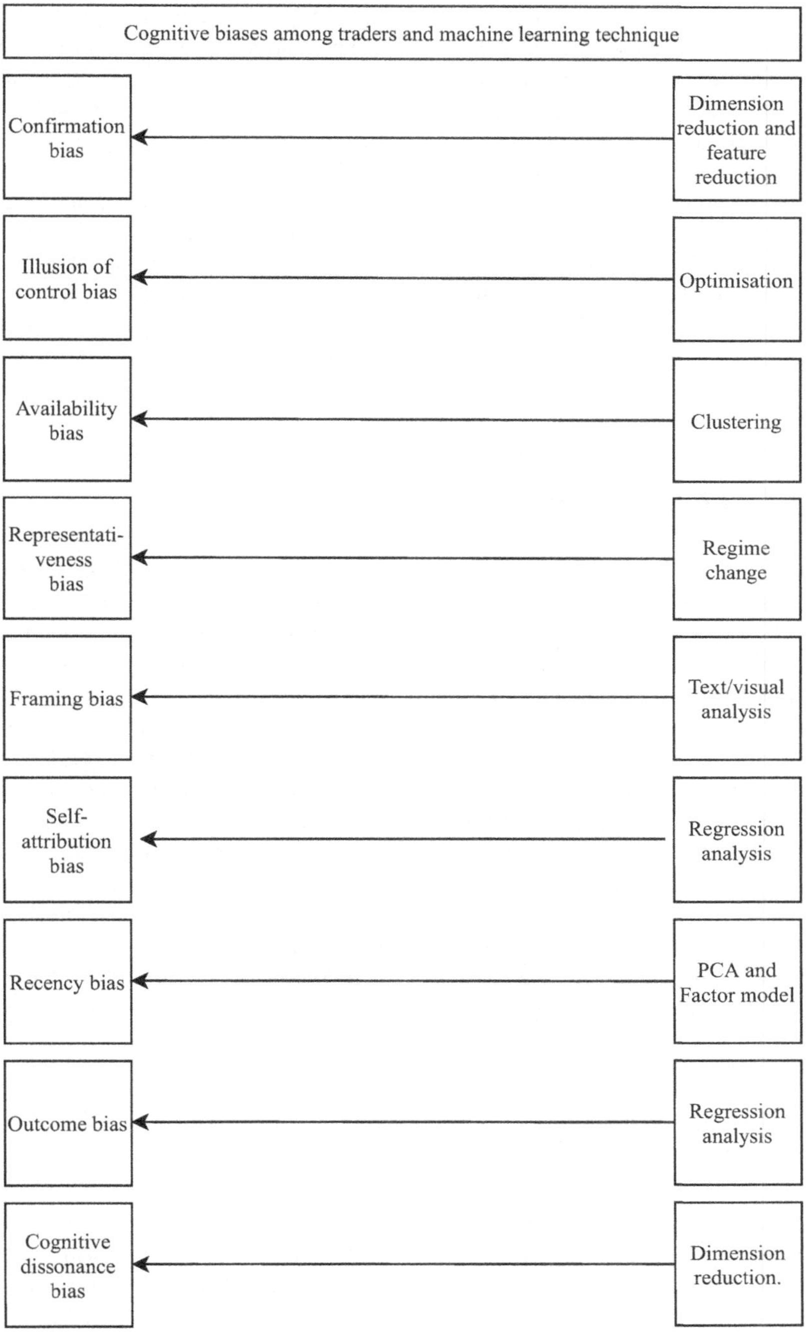

Fig. 16.1 Application of machine learning techniques to overcome traders' behavioral biases.

weight to the information that supports their buy or sell recommendation and ignore the relevant (material) information that contradicts their existing beliefs. This bias tends to sub-optimal decision-making by traders and results in lower performance. Traders at the arrival of new information can use unsupervised machine learning techniques such as dimension reduction and feature extraction techniques to make optimum use of the newly arrived information. The application of the above-mentioned techniques will help traders to focus only on the relevant information and such focus will help to improve their performance. Information is vastly available in the stock market and traders need to filter only the relevant information. Unsupervised machine learning provides helpful tools to handle vast information. The use of unsupervised machine learning techniques helps traders to focus on the vital pieces of information and objectively ignore much of the redundant information.

16.1.2 Illusion of Control Bias

Traders often overestimate their ability to control the events and confuse their commitment to task with the illusion of control. Better ability and dedication for a buy or sell recommendation do not lead traders to control the uncertain or probabilistic events. The decisions of traders should be based on rigorous analysis and not on their dedication or commitment to a transaction. The presence of bias of illusion of control leads traders to not focus on the simulated results or/and sensitivity analysis and hence the performance of traders gets negatively affected. Traders provide too much extra weight to their analysis and their confidence in their analysis is not optimum. To reduce the impact of the illusion of control bias, traders can apply machine learning tools such as the optimization techniques for a task from reinforcement learning. The application of optimization techniques can be efficiently done to minimize trading costs. Traders can reduce the subjectivity involved in their decisions with the help of optimization techniques. The noise in the estimated expected return or volatility of the market is also affected by the illusion of control bias. The noise in the forecasts of estimates of traders can also be reduced by using suitable machine learning techniques.

16.1.3 Availability Bias

Availability bias deals with the tendency of individuals to assign more weights to the readily available information and assign less significance to information that is more abstract. Traders prefer information that is readily available because new information is costly to acquire in the financial markets [13, 14]. Decisions based on ready information and with no or less focus on salient information are sub-optimal in nature and deteriorate traders' performance. To focus on the information with similar significance and content, traders can use the clustering techniques of unsupervised machine learning. Besides this, to focus on the required information especially the abstract information, traders can also use hierarchical clustering. The information with similar features can be effectively grouped using the above techniques. Clustering helps the traders to label the information based on the significance and a tree structure or dendrogram helps the traders to make sound decisions.

16.1.4 Representativeness Bias

Individuals make forecasts based on the heuristics and then classify the uncertain events to one of the existing classes of events. The arrival of new information is not always related to an existing or past event; the shortcut to relate the arrival of new information to an existing one is harmful to optimal decision-making. To reduce the negative consequences of the representativeness bias, traders can use regime change unsupervised machine learning tools from representation learning as such tools help the traders classify the information in a new regime. The focus on a new regime will further help the traders to analyze the information with a new perspective highlighting a new set of characteristics to make informed decisions. Decision-making and forecasts of traders will certainly improve with the use of the regime change technique, as traders will classify information either in a new class or in a more relevant existing class. The regime change technique is promising for traders when doing sensitivity or scenario analysis. The bias of representativeness can be mitigated using the ability of the above-stated technique to add a new perspective to the existing problem.

16.1.5 Framing Bias

When individuals change decisions based on how the information is presented, it is called framing bias. Traders make faulty decisions based on the way a change in policy or key rates are presented to the traders. The framing of words alters the decision-making process and deviates traders' from optimal decision-making. Proper text or visual recognition by traders will help them resolve the consequences of framing bias. Traders should not be making decisions based on the way the information is presented and to help them in mitigating the impact of this framing bias, the proper text recognition, and analysis techniques yield positive outcomes. Proper text or visual analysis may also highlight one of the key missing pieces of information involved in the decision-making. The ability of traders to accept views from a diverse group of participants and evaluating the feedback of the past decisions will surely help traders to enhance their performance.

16.1.6 Self-Attribution Bias

When individuals take the credit for success and avoid taking the responsibility for failure, they exhibit self-attribution bias. Traders confuse short-term success and long-term success; they often take credit for short-term success and blame uncertainty of macroeconomic variables for the long-term failures. The use of a multiple regression model to evaluate the ex-post performance of a trading strategy will help the trader to reduce the impact of self-attribution bias. A trader can run a regression model to find the significant variable used in the prior successful strategies and it will help him/her to do a more objective analysis. The significance of variables used in the prior strategies or variables based on the intuition of the trader can also be analyzed using regression analysis. Regression analysis helps the trader to identify the relevance of the variable based on historical information and this assists the trader to not relate the random variable with their success of the strategy.

16.1.7 Recency Bias

Individuals often emphasize recent events to take decisions and ignore the requirement of a rigorous process to make decisions. At times, traders do not allocate their funds as per the investment process and policy, instead focus on the recent superior performance of events to take positions in the market. Actions like these result in the deteriorated performance of events and such deterioration of performance can be handled with suitable machine learning techniques. Factor models and principal component analysis (PCA) tools of unsupervised machine learning assist traders to put emphasis not only on the recent events but also on the prior events. The required dimensionality reduction can be applied with the help of machine learning techniques to assist the traders to consider only relevant aspects of information. A thorough analysis of all or most of the past strategies by the trader will help the trader reduce the influence of the recent events on the decision-making. Traders can also use the regime change method to focus on the new scenario from a different perspective. Regime change also helps the traders to get rid of the recent events and highlights new parts of the new scenario, which are potentially different from the recent experience of the trader.

16.1.8 Outcome Bias

When individuals make decisions based on the past outcomes of the events and not the process, which describes the rationale for such an outcome, they are said to suffer from outcome bias. Traders take future bets based on the outcome of present bets ignoring the reasons for such outcomes. Outcomes might be derived from factors that are not studied by the trader and noise plays a crucial role in scenarios where traders do not account for all the relevant variables in the decision-making process. Multiple regression analysis is useful for traders to reduce the negative consequences of the outcome bias. The regression model helps the trader to focus on the predictors and such focus helps the traders to eliminate undue emphasis on the outcome of the event. Illogical reasoning drives the decisions based on the outcome and those decisions deteriorate the performance of traders.

16.1.9 Cognitive Dissonance Bias

Individuals experience cognitive dissonance when the newly arrived information contradicts the existing knowledge about an event. Traders often experience cognitive dissonance when the strong opinion about a security selection needs to be updated after the arrival of negative information but the trader does not update the decision because the negative information contradicts with the trader's existing favorable information about the security. Analysis of the newly arrived information through the regression models helps the traders to quantify the significance of the new information (variables). Besides, dimension reduction techniques can also help the trader to focus on the most vital piece of information (variable). Regime change also helps the traders to focus on the new information with new scenarios. The focus on contradictory information helps traders to evaluate their buy or sell recommendation. The status quo needs to be updated, if required, with the new information and machine learning techniques help the traders to make better-informed decisions.

The above section lists various cognitive biases present in the behavior of traders and helps the traders by focusing on relevant machine learning techniques to make better decisions. The use of supervised and unsupervised machine learning techniques ensures the traders of sound decision-making and better financial performance. Also, traders suffer from the presence of emotional biases such as loss aversion, optimism, overconfidence, status quo, endowment, and regret aversion. Emotional biases are difficult to overcome or treat and [15] highlight the role of behavioral biases in decision-making. Emotional biases can only be adopted and not moderated. Hence, the focus of the present chapter is only on the cognitive biases of traders as these biases can be adapted. The management of cognitive biases helps traders to increase financial performance and to improve their decision-making.

Some studies in the literature have explored the role of analytics in different areas of finance such as return estimations, forecasts of financial variables, and improved predictions. Chen et al. [16] conducted a study on the use of big data analytics (BDA) and value creation; researchers showed theoretical implications based on BDA and management. This is a new dimension of integration of BDA and it shows how management can use BDA to increase value, also showing the promising role of analytics in fields such as supply chain. G. Chen et al. [17]

studied the trading performance with investors' bias to draw inference from China. Researchers documented the pattern in the behavior of investors and showed how investors deviate from optimal decision-making. The study presents the need for the recognition of such biases and analytics provides a framework to achieve the desired goal of recognition of behavioral biases and their treatment. This chapter recognizes the presence of such biases and highlights the role of analytics to prevent decision-making mistakes.

Boyacioglu [18] applied algorithms in the stock market predictions using adaptive network-based fuzzy inference system (ANFIS) and showed that algorithms help in better predictions. If analytics help in better predictions, then the use of analytics in the traders' behavior will yield more promising results. If traders' behavior can be adapted to an optimal level, then the traders will form predictions that are more accurate and such accurate predictions will result in better results. Seddon [19] applied data analytics tools in the prediction of daily stock returns and documented the evidence of better forecasts with few inputs to the model. They used analytics tools to present the evidence of estimation of down movements with more accuracy and such forecasts can be used in place of value-at-risk (VaR) and similar risk measures.

Nann [20] studied the role of analytics in high-frequency trading (HFTs). The study documented the significant difference between the HFTs and low-frequency trading (LFTs) in terms of competitive asymmetries. The present chapter helps traders not only with regard to HFTs but also LFTs and acts as guide to use analytics to minimize their behavioral errors and focus on their performance. [21] applied predictive analytics on the public data of the stock market and stated the results using a trading model. The model using predictive analytics helped the researchers to earn a positive return higher than the benchmark. The above studies have shown the significance of analytics in diverse fields in addition to stock markets.

16.2 Concluding Remarks

The chapter highlights various cognitive biases documented in the behavioral corporate finance literature among individuals and applies the framework of machine learning techniques to assist the biases. The behavioral biases of confirmation, the illusion of control, availability, representativeness, framing, self-attribution, recency, outcome, and cognitive dissonance are stated in the present chapter. Traders make certain decisions concerning forecasts such as the bid price, ask price, expected return, expected volatility, a continuation of certain patterns in the market, and trading costs. The biased behavior of traders deviates their decision-making from optimum decisions to sub-optimum decisions. The awareness about behavioral biases and the application of suitable machine learning tools to handle such biases will certainly help the trader to make informed decisions. The assumption of rationality about the behavior or decisions of traders is challenged in the present chapter, and relevant literature from the field of finance and psychology is presented.

The biased behavior due to the presence of cognitive biases can be adapted using machine learning techniques such as cluster analysis, regime change, regression analysis, text recognition, visual analysis, classification methods, feature extraction, dimension reduction, factor models, and principal component analysis. The focus on different machine learning tools also states that no single machine learning technique is universally applicable and to tackle different cognitive biases, different techniques are required. The present chapter is useful for traders, analysts, and academicians interested in the integration of finance, psychology, and machine learning. With the advancement of research in behavioral finance and machine learning, better techniques to mitigate the impact of cognitive biases on the decision-making will result in the superior financial performance of traders. We hope that the focus on machine learning techniques with regard to the cognitive biases of traders helps them in sound decision-making and they benefit in terms of superior performance by integrating the recommended machine learning to mitigate the impact of cognitive-behavioral biases.

References

1. H.M. Shefrin, R.H. Thaler, The behavioral life-cycle hypothesis. *Econ. Inq.***26**(4), 609–643 (1988). https://doi.org/10.1111/j.1465-7295.1988.tb01520.x
2. D.F. Costa, Carvalho, F. de M., & Moreira, B. C. de M., Behavioral economics and behavioral finance: a bibliometric analysis of the scientific fields. *J. Econ. Surv.***33**(1), 3–24 (2019). https://doi.org/10.1111/joes.12262
3. V. Ramiah, X. Xu, I.A. Moosa, Neoclassical finance, behavioral finance, and noise traders: A review and assessment of the literature, in *International Review of Financial Analysis*, (Elsevier, Amsterdam, 2015), pp. 89–100. https://doi.org/10.1016/j.irfa.2015.05.021
4. N. Barberis, R. Thaler, A survey of behavioral finance, in *Handbook of the Economics of Finance*, (Elsevier, Amsterdam, 2003), pp. 1053–1128
5. R.H. Thaler, The end of behavioral finance. Financ. Anal. J. **55**(6), 12–17 (1999). https://doi.org/10.2469/faj.v55.n6.2310
6. R.H. Thaler, Behavioral economics: Past, present, and future. *Am. Econ. Rev.***106**(7), 1577–1600 (2016). https://doi.org/10.1257/aer.106.7.1577
7. H. Shefrin, M. Statman, The disposition to sell winners too early and ride losers too long: Theory and evidence. *J. Financ.***40**(3), 777–790 (1985). https://doi.org/10.1111/j.1540-6261.1985.tb05002.x
8. H. Shefrin, M. Statman, Behavioral capital asset pricing theory. *J, Finan, Quantit. Anal.***29**(3), 323 (1994). https://doi.org/10.2307/2331334
9. H. Shefrin, M. Statman, Behavioral portfolio theory. *J, Finan, Quantit. Anal.***35**(2), 127 (2000). https://doi.org/10.2307/2676187
10. H. Shefrin, Behavioral corporate finance. *J. Appl. Corp. Financ.***14**(3), 113–126 (2001). https://doi.org/10.1111/j.1745-6622.2001.tb00443.x
11. Y. Zhang, P.E. Swanson, Are day traders bias-free?-evidence from internet stock message boards. *J. Econ. Finan.***34**(1), 96–112 (2010). https://doi.org/10.1007/s12197-008-9063-1
12. A. Cafferata, F. Tramontana, A financial market model with confirmation bias. Struct. Change Econ. Dyn. **51**, 252–259 (2019). https://doi.org/10.1016/j.strueco.2019.08.004
13. S. Grossman, On the efficiency of competitive stock markets where trades have diverse information. J. Financ. ***31***(2), 573–585 (1976). https://doi.org/10.1111/j.1540-6261.1976.tb01907.x

14. R.A. Ippolito, efficiency with costly information: a study of mutual fund performance 1965–1984. Q. J. Econ. **104**(1), 1 (1989). https://doi.org/10.2307/2937832
15. A. Chandra, Decision-making in the stock market: incorporating psychology with finance, in National Conference on Forecasting Financial Markets of India, 2008, pp. 1–28. http://ssrn.com/abstract=1501721
16. D.Q. Chen, D.S. Preston, M. Swink, How the use of big data analytics affects value creation in supply chain management. J. Manage. Inform. Syst. ***32***(4), 4–39 (2015). https://doi.org/10.1080/07421222.2015.1138364
17. G. Chen, K.A. Kim, J.R. Nofsinger, O.M. Rui, Trading performance, disposition effect, overconfidence, representativeness bias, and experience of emerging market investors. J. Behav. Decis. Mak. ***20***(4), 425–451 (2007). https://doi.org/10.1002/bdm.561
18. M.A. Boyacioglu, D. Avci, An adaptive network-based fuzzy inference system (ANFIS) for the prediction of stock market return: The case of the Istanbul stock exchange. Exp. Syst. Appl. ***37***(12), 7908–7912 (2010). https://doi.org/10.1016/j.eswa.2010.04.045
19. A. Oztekin, R. Kizilaslan, S. Freund, A. Iseri, A data analytic approach to forecasting daily stock returns in an emerging market. *European J. Operation. Res.* ***253***(3), 697–710 (2016). https://doi.org/10.1016/j.ejor.2016.02.056
20. J.J.J.M. Seddon, W.L. Currie, A model for unpacking big data analytics in high-frequency trading. *J. Bus. Res.* ***70***, 300–307 (2017). https://doi.org/10.1016/j.jbusres.2016.08.003
21. S. Nann, J. Krauss, D. Schoder, Predictive analytics on public data-the case of stock markets. ECIS **2013**, 102 (2013)

Chapter 17
Real-Time Data Visualization Using Business Intelligence Techniques in Small and Medium Enterprises for Making a Faster Decision on Sales Data

Owes Quruny Shubho, Zerin Nasrin Tumpa, Walid Ibna Rakib Dipto, and Md Rasel Alam

17.1 Introduction

Data analysis to visualize is a process of improving the performance of business systems and achieve its goal also to compete with others. But it is a major challenge to capture consecutive changes in data when client wants change frequently. Furthermore, matching the need for data collection with the storage requirements of a data warehouse is a difficult task. Therefore, we support the access for data collected instantly and more frequently with fastest ad hoc queries to collect data more rapidly. Likewise, our model makes it happen to analyze visualization in quite a simple way. It is a process of input data collected from the user into an application, transforming information into a visual form enabling the viewer to observe the downward trend and hike uptrend. However this model helps to deliver on time and give the best output with various mediums such as web view, mobile, or desktop.

This model can provide results in any situation when there is a data update. The user can experience an outstanding competence business intelligence system which makes easier than others.

17.2 Business Intelligence Literature Review

Business intelligence is a way of gaining advantage from business using data. This data can be user's information, stock information, sales report, or any source that

O. Q. Shubho (✉) · Z. N. Tumpa · W. I. R. Dipto · M. R. Alam
Department of Computer Science and Engineering, Daffodil International University, Dhaka, Bangladesh
e-mail: quruny15-8308@diu.edu.bd; walid15-8362@diu.edu.bd; rasel15-8310@diu.edu.bd

P. M. Jeyanthi et al. (eds.), *Decision Intelligence Analytics and the Implementation of Strategic Business Management*, EAI/Springer Innovations in Communication and Computing, https://doi.org/10.1007/978-3-030-82763-2_17

Table 17.1 Table of facts about business intelligence

Facts	Percentage
Manager thinks that they have enough information at their workplace	61%
Think and believe they will get worse on their data	80%
Managerial staff thinks about information overload and information garbage for that they avoid data for real-time decision making	50%
Managers store their information for future use, rather than using that information for present analysis	84%
Information gathering and analysis cost is outweighed its value	60%
Executives companies cannot have real-time information so they cannot take decision rapidly	77%

is related to its business. Data may also come from a variety of sources, including client business transactions, retail, RFID tags, email, and query logs [1]. Business intelligence extracts information from a significant quantity of data and converts it into knowledge, which is used by decision support systems. Business intelligence is a mass effective way to make a data-driven decision [2]. Business intelligence visualizes data and gives us a visual look of data that can be easily understood [3]. According to the authors of the publications "Business Intelligence Using Data Mining Techniques and Business Analytics" and "Implementation Benefit to Business Intelligence Data Mining Techniques," today's businesses are suffering from information overload [4, 5] (Table 17.1).

According to those facts we assume that we have information but we cannot deal with that and business decision maker cannot handle or manage that information in an efficient way. So the data and information stay unused. We can see there is a gap for doing those data processing. And day by day the data volume also increasing so fast. Data volume is increasing at the speed of 44X in the next decade it will reach 800 K petabytes to 35 zettabytes [6].

Previously business data or sales data were analyzed by statistical techniques, but nowadays it is outdated. Only statistical techniques are not enough for real-time business decision and data analysis. We need to mine the data related to our business. Business intelligence techniques help us to do that [7].

17.3 Methodology

In traditional business intelligence architecture observed that mobile sync is not established and all data comes from the central data warehouse [1]. From that perspective our proposed model demonstrates the process of data collection for visualization. While the process is complete with seven different segments wherein the initial stage data is collected from the input field (internal or external users) and store in a local database as offline [8]. When it is linked to the internet, however, the data is stored in a special database called Firebase cloud storage [9]. This data

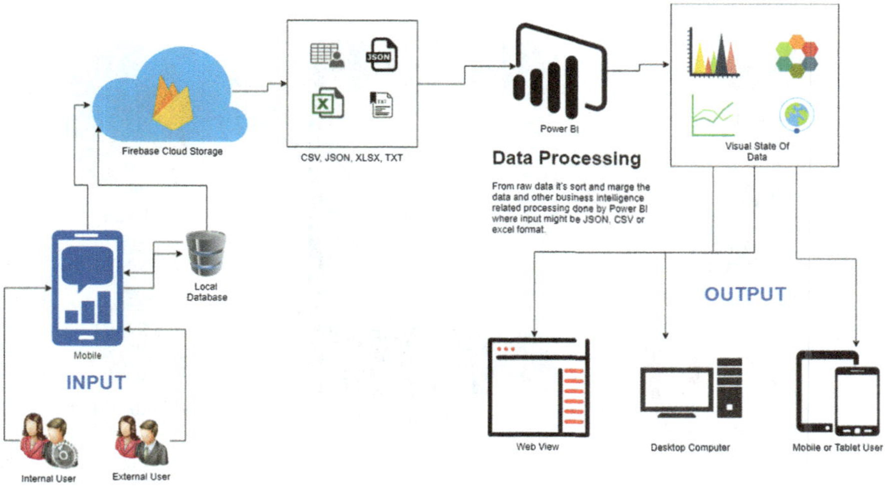

Fig. 17.1 Figure of proposed model "real-time data visualization using business intelligence techniques in small and medium enterprise for making a faster decision on sales data"

will be count as raw data, this data are going to short and merge in our next stage of data processing which is very important for our model. As a result, raw data is processed by a business analytics service, which includes interactive visualizations and business intelligence capabilities with an interface that is easy enough for end-users to build intelligent reports and dashboards [10]. Finally, in the last process as visualization, it gives output through electronic devices such as mobile, desktop, and web view (Fig. 17.1).

17.3.1 Data Collection Phase

There is a mobile interface it will be a mobile application (Android-based OS) which is connected to Firebase throw the internet. In the input phase that mobile application also works in offline mode that time all data are saved in mobile local storage. After connected with internet all data automatically sync with Firebase. The mobile application of this model has two user access, one internal staff of the business (regional managers) and external staff of the business (general sales staff). They also have a login Id and password to access their input dashboard. Security maintains with two-factor authentication (2FA) using Account Kit [11]. Various sources of data can be directly applied to power bi. It supports almost all common database and format used in business purpose (Fig. 17.2).

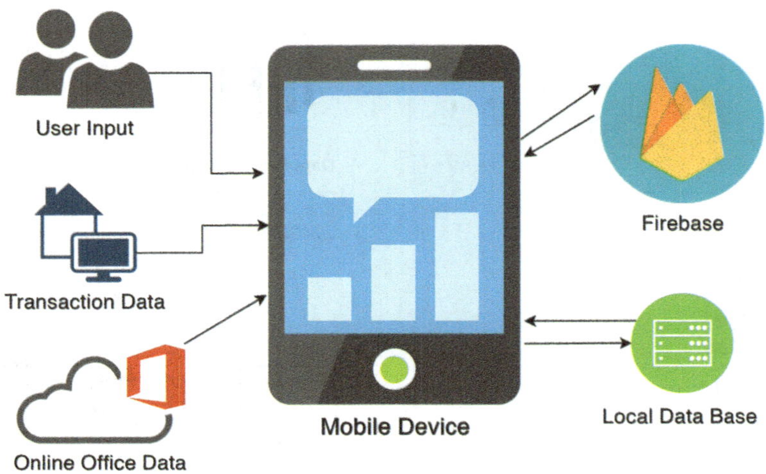

Fig. 17.2 Figure of data collection phase

17.3.2 Data Store and Processing Phase

Data is stored as JSON [12] and synchronized in real-time to every connected client. We also store raw file and formatted file like excel. The data can be accessed by online application or third part application throw API. Here is the most important part of our model Power BI [13]. We can freely download a desktop version of Power BI desktop without any cost. It is freeware. In the under layer of power, business intelligence used an in-memory analytical engine and columnar database that tabular data store structures which used power pivot [14]. Also Power BI includes Power Query that is a total package of extract, transform, and load which can turn data to usable and ready to work [15]. Using DAX (Data Analysis Expressions) [16] Power BI produced extended and expressed model of data.

17.3.3 Dataset

The proposed model used an excel file that contains five tables.

Manger's Table The business is running on several suburbs where they have few managers who manage that suburbs. In the manager's table it has some information such as managers name, their maintain suburbs and suburbs postcode.

Buyers' Table There are certain customers in the buyers table who purchase a few categories of items. This figure displays the buyers' names as well as their purchasing categories.

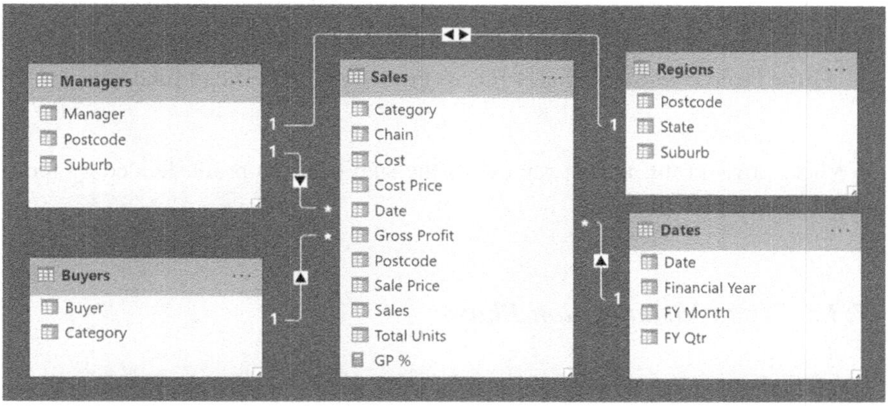

Fig. 17.3 Data fields and data relations

Regions Table Suburbs are located in several states and areas. The status of suburbs and their postcodes are shown in a figure.

Dates Table The whole business is maintain a financial year and it have financial year quarter. In the dates table financial year quarter and its financial year in months mentions.

Sales Table Sales table shows the chain name, postcode of that chain, category that sold, total units of sales, sales price, and cost price with dates.

Using Power Query related data type fields are connected [17]. These tables have a relation to their common field of data (Fig. 17.3).

17.3.4 Equations

To archive the main result in our Power BI we use some DAX formula that helps us to get total sales amount, cost, gross profit, and gross profit in percentage.

$$\text{Total Sales} = [\text{Sales Price}] * [\text{Total Unit}] \tag{17.1}$$

In Power Query mode of power bi, there is a new column added and its total sales which is multiplying of Sale Price column and Total Unit column.

$$\text{Cost} = [\text{Cost Price}] * [\text{Total Unit}] \tag{17.2}$$

Cost column is a result of multiplying cost price column and total unit column.

$$\text{Gross Profit} = [\text{Total Sales}] - [\text{Cost}] \tag{17.3}$$

Gross profit is a profit that when total sales subtract by cost.

$$\text{Gross Profit\%} = \text{SUM (Sales [Gross Profit]) /SUM (Sales [Total Sales])} \tag{17.4}$$

Where gross profit percent represents the sum of gross profit divided by totals sales.

17.3.5 Data Visualization Phase

Using a variety of procedures and process ideas, such as plotting data and merging data. Various results were displayed on a summary dashboard [18]. Which is more than a data visualization it's related to business and shows some insight of business information that's can't show on raw or other media of data. Data visualization is more useful than just a sales report since it allows you to see company data via numerous charts and infographics (Fig. 17.4).

State Filter Section (1) This state filter section has a strong filter capability to filter data by state. According to the state all related data also automatically changed. State filter will show the sales report and other key performance indicator by only selected state.

Key Performance Indicator (2) At a glance, this board showed the total sales with background timeline chart. That might reveal the status of the company's sales,

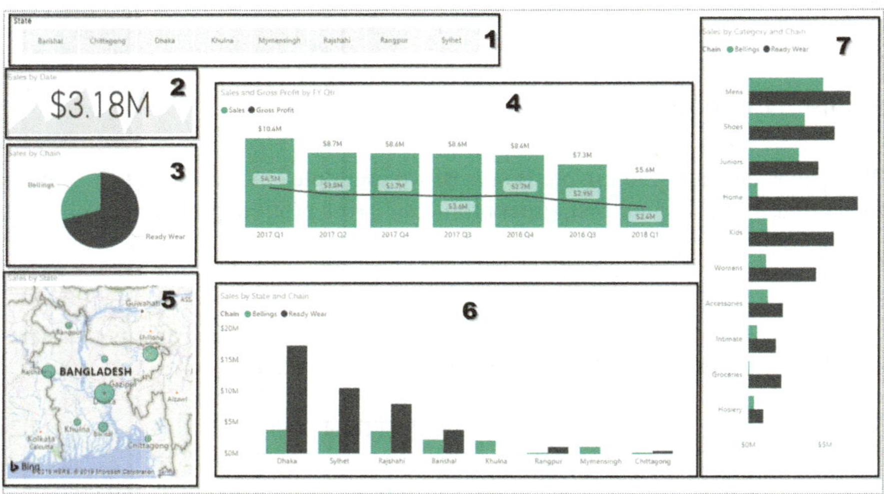

Fig. 17.4 Figure for dashboard of proposed model

which can assist in the management of a specific section that is directly affecting overall sales. This is very significant for our company leaders when it comes to making decisions [19].

Pie Chart (3) The Pie chart shows the part of sales by the chain where business decision-maker can see the portion which is larger and which is smaller.

Line and Stacked Column Chart (4) Using Sparkline, a stacked column shows the increase in sales over the course of four financial quarters, and a line inside the stacked column shows the increase in gross profit over the course of four financial quarters.

Map (5) It displays the sales strength of each state inside a map format. This is an overview of the sales by state. This is a state-by-state breakdown of sales. Geo location wise sales indicator help a lot for making changes on product supply and ready stock for next selling sessions.

Clustered Column Chart (6) Clustered column chart is a combination of state sales data with the chain. It shows the total sales by the state with the chain.

Clustered Bar Chart (7) Clustered bar chart is also known as a group bar chart that can show some group of data at a single view of the chart. Our sales category and chain data are combined in the clustered bar chart.

17.3.6 Scatter Chart

A scatter chart (Plot) is a mathematical representation that uses Cartesian coordinate to present a set of data to a visual coordinate. It uses linear regression that maintains a finite time of data to view the exact result [20].

Here our X-axis presents the sales values, Y-axis growth profit in percent and the size of the plot is gross profit and the play button is the financial quarter. Proposed model gives us state filter to find sales by states, main KPI to indicate sales state, sales and gross profit by any financial year quarters, category and chain wise sales (Fig. 17.5).

17.3.7 Sparkline

Proposed model produces few Sparkline which shows the sales trends by different categories. For example, state-by-state sales, or a comparison of total sales price vs. total cost price. The sparkline indicates that the sales price trend begins at a low point, whereas the cost price trend begins at a high point. In the financial Sparkline

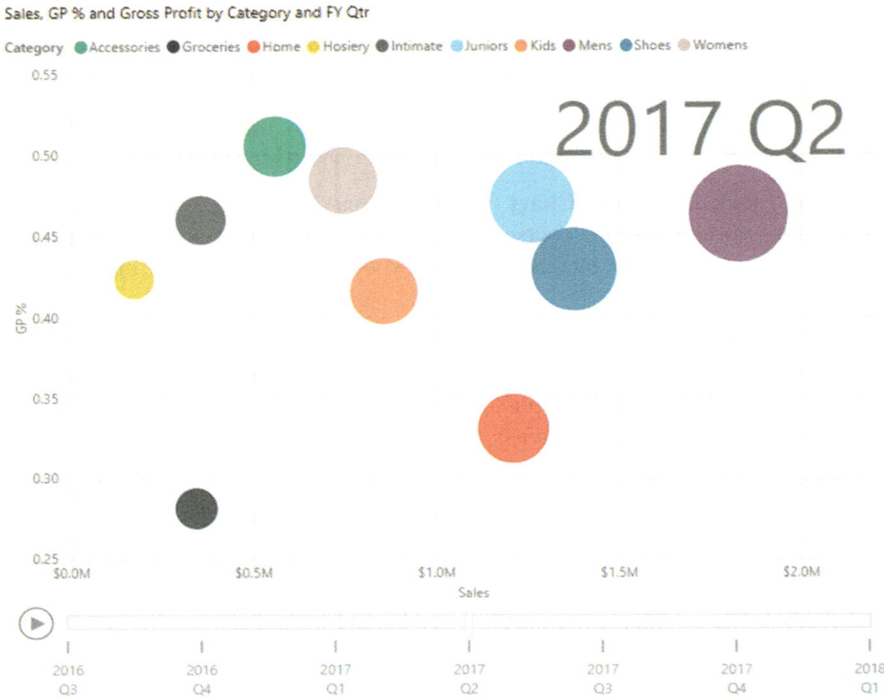

Fig. 17.5 Figure of scatter chart

we saw that first year is not fluctuated but mid-year is so fluctuated and last year shows uptrend. Gross profit Sparkline shows trends that start with down and now indicate uptrend and also show lowest and highest gross profit which can help to find peak and off-peak sales sessions (Fig. 17.6).

17.3.8 Cross Platform and Device View

This dashboard has the capability that can view online and there also has a power BI mobile app that can show the published dashboard on the mobile. Using Web View and iframe this dashboard can viewable on any webpage and device that support Web View and iframe technology [21, 22].

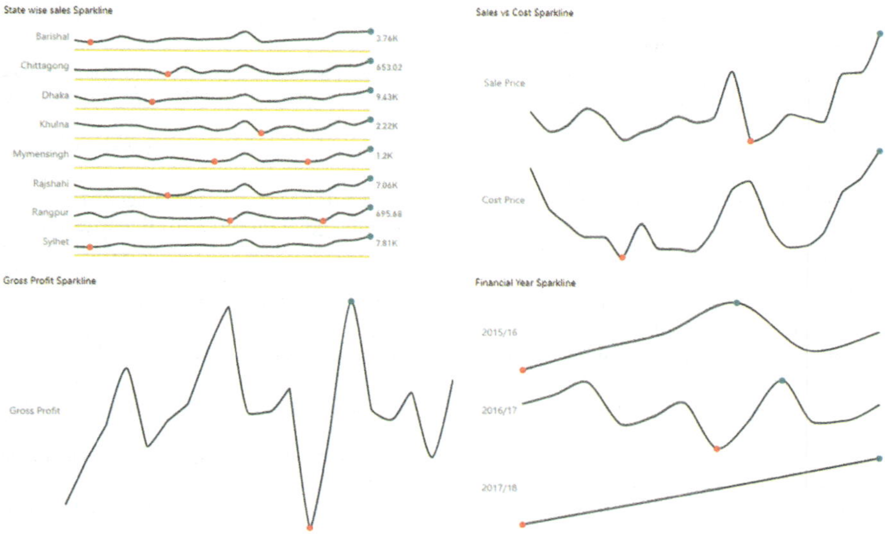

Fig. 17.6 Figure of Sparkline

17.4 Conclusion

In conclusion, BI has a big impact on the IT industry. Every corporate organization manages insights into their data, strategies, profits, and performances, thus BI helps them to make significant and complex decisions. This paper represents important facts about data collection, storage, process, and visualization output. The solution mainly based on power BI insight where table relationship and data-driven culture by enabling everyone to turn data into insightful visualizations they can use to make business decisions quickly and confidently. From ordinary sales data by using DAX and other business intelligence technique our proposed model is a small working framework that actually helps to understand the sales verdict.

According to our suggested approach, we take data from internal and external sources, analyze it in near real-time, and provide an immediate cross-dependent overview that may display overall sales KPIs, highlighted sales locations by geolocation, and Sparkline can assist forecast sales trends over time.

References

1. S. Chaudhuri, An overview of business intelligence technology. Commun. ACM **54**(8), 88 (2011)
2. OLAP.com, What is Business Intelligence? BI Definition, OLAP.com, 2019. https://olap.com/learn-bi-olap/olap-bi-definitions/business-intelligence/. Accessed 18 November 2019

3. J. Ranjan, Business intelligence: concepts, components, techniques and benefits. J. Theor. Appl. Inf. Technol. **2009**, 60 (2009)
4. B.K. Mishra, D. Hazra, K. Tarannum, M. Kumar, Business intelligence using data mining techniques and business analytics, in 5th International Conference on System Modeling & Advancement in Research Trends, SMART-2016, Moradabad, India, 2016
5. H. Singh, Implementation benefit to business intelligence using data mining techniques. Int. J. Comput. Business Res. **2012**, 2229–6166 (2012)
6. Executive summary, data growth, business opportunities, and the IT imperatives, EMC Dell, April, 2014. https://www.emc.com/leadership/digital-universe/2014iview/executive-summary.htm. Accessed 18 Nov 2019
7. Hiba Alsghaier, et al., The impact of big data analytics on business competitiveness, in Proceedings of the New Trends in Information Technology (NTIT-2017), The University of Jordan, Amman, Jordan, 2017
8. M. Al Hasan, An economic and modern business intelligence solution for textile. Glob. J. Comp. Sci Technol. **18**(2), 12 (2018)
9. P.S. Chunnu Khawas, Application of firebase in android app development. Int. J. Comput. Appl. **179**(46), 49–53 (2018)
10. H. Cheng, Y.C. Lu, C. Sheu, An ontology-based business intelligence application in a financial knowledge management system. Expert Syst. Appl. **36**, 3614 (2009)
11. A. Amin, I.u. Haq, M. Nazir, Two factor authentication. Int. J. Comput. Sci. Mob. Comput. **6**(7), 5–8 (2017)
12. J. Freeman, What is JSON? A better format for data exchange, InfoWorld, 25 October 2019. https://www.infoworld.com/article/3222851/what-is-json-a-better-format-for-data-exchange.html. Accessed 18 Nov 2019
13. Vijay Krishnan, S Bharanidharan, G Krishnamoorthy, Research data analysis with power BI, in 11th International CALIBER-2017, Anna University, Chennai, Tamil Nadu, 2017
14. Wikipedia, Power Pivot, Wikipedia, 30 May 2019. https://en.wikipedia.org/wiki/Power_Pivot. Accessed 18 Nov 2019
15. K. Gowthami, M.R.P. Kumar, Study on business intelligence tools for enterprise dashboard development. Int. Res. J. Eng. Technol. **4**(4), 2395–0072 (2017)
16. Zhong, Power BI: What is DAX? And why you should (or should not) learn it, 2018. https://towardsdatascience.com/power-bi-what-is-dax-and-why-you-should-or-should-not-learn-it-1581aadbb038. Accessed 18 Nov 2019
17. J. Acampora, Power query overview: an introduction to Excel's most powerful data tool, Excelcampus, 20 March 2015. https://www.excelcampus.com/power-tools/power-query-overview/. Accessed 18 Nov 2019
18. M. Rouse, Business intelligence dashboard, Search Business Analytics, 2018. https://searchbusinessanalytics.techtarget.com/definition/business-intelligence-dashboard. Accessed 18 Nov 2019
19. I. Bhatti et al., *The key performance indicators (KPIs) and their impacton overall organizational performance* (Springer, Dordrecht, 2013). https://doi.org/10.1007/s11135-013-9945-y
20. D.A. Keima, M.C. Haob, et al., Generalized scatter plots. Inform. Visualiz. **9**(4), 301–311 (2010)
21. H. Paul, What a webview is and how to test it, DZone, 9 October 2018. https://dzone.com/articles/what-is-a-webview-and-how-to-test-it. Accessed 18 Nov 2019
22. M. Rouse, IFrame (Inline Frame), https://whatis.techtarget.com/, 2015. https://whatis.techtarget.com/definition/IFrame-Inline-Frame. Accessed 18 Nov 2019

Chapter 18
To Invest or Not to Invest? A Case Study with Decision Analytics on Japanese Yen

Swamynathan Ramakrishnan, Sredharran Sampath, Prannav Srikanth, and A. Mansurali

18.1 Introduction to Foreign Exchange Market

Foreign exchange market stands atop as the biggest and easy to convert liquid cash markets in the world with the amount being traded in trillions. It is open 24 h a day, 7 days a week, this market is accessible to anyone from public banks and multinational corporations to the individual trader. According to McFarlin [1], the everyday average turnover is around $4 trillion. He has identified that swapping in currency exchange points out the exchanging one currency in return of another either to safeguard imminent risk due to vulnerability in a particular currency or to gain from rate of exchange fluctuations.

As defined by Rosenberg [2], the legal tender can be exchanged either with spot rate, forwards rate, exchange traded funds, or options Market. It has been found that recently the spot rate dominates price determination.

Individuals and experts still have their doubts about the trends of the currency market and whether certain charts could be used to foretell future movements. The gainfulness of trading the currency in technical way has been reducing considering the late 1980s [3]. There is proof that there are lesser gains from trend-trading than

S. Ramakrishnan (✉)
Amity University, Dubai, UAE
e-mail: sramakrishnan@amityuniversity.ae

S. Sampath
Aspire Training Consultancy, FZE, Fujairah, UAE

P. Srikanth
Cornerstone College of International Studies, Sharjah, UAE

A. Mansurali
PSGIM, PSG College of Technology, Coimbatore, India
e-mail: mansurali@psgim.ac.in

P. M. Jeyanthi et al. (eds.), *Decision Intelligence Analytics and the Implementation of Strategic Business Management*, EAI/Springer Innovations in Communication and Computing, https://doi.org/10.1007/978-3-030-82763-2_18

done previously, yet the recent trading profits for non-native currencies are still affordable [4].

Steve Nison [5] in his book on Japanese candlestick techniques has made a mention that Japanese candlestick technique is an adaptable tool that can be combined with any other scientific tool and will help upgrade any technician's market analysis. Northcott [6] gives a caution that candlesticks must not be used in solo to make an exchange determination. They do not highlight sufficiently about the rest of the price activity, and their explanation repeatedly hangs on the shift they are in. One must regulate the comprehensive market position using traditional technical pointers prior to entering into a trade. Candlesticks work excellent at stipulating turnaround positions when the price is overestimated or overvalued, in which case they could aid with the scheduling of one's entry. In these circumstances, a doji candle shows that none is administering, neither bulls nor bears, so the swing is unbiased. Individual candlesticks such as the hanging man and hammer emergence exhibit an abundance of information and can show the likelihood of a one-day shift, yet there is also the chance that they could simply be deviations, so designs made of various candlesticks can present proof that the shift is real.

18.2 Japan and Its PPP

Using PPP (purchasing power parity) as a parameter, after China, the USA, and India, Japan stands fourth in the size of economy as measured. Gustave Cassel [7] has defined PPP as the value of a legal tender that is decided by the acquiring power of goods and services it initiates. However, Japan has been agonizingly going through deflation and slow growth since the 1990s. Being part of the G7 has ensured Japan a strong economic recovery in the year 2020. By the fourth quarter, the GDP of Japan has been down by just 1.2% compared to the previous year, mainly due to a recovery in consumer spend and abundant foreign demand. Compared to the USA and EU, where the real GDP has been down by 2.4% and 4.8% individually. One among the many factors that has contributed to this situation is that support for Japan's economy has come from outside Japan. Exports to the Asian countries including China have been strong. Japan has posted a double-digit export growth when compared with the growth achieved by countries such as India, Thailand, and Malaysia. The USA has indirectly supported Japan's strong export growth as well. It acts as the final buyer of exports from Japan. Japan's abundant exports include automobiles, valuable metals, machinery parts, electrical and electronic components. However, with the world momentum shifting in favor of electric vehicles to counter the global warming, which might hurt Japan's economy. The reason being electric vehicles make use of parts that are less by one-third in comparison to oil driven vehicles.

Japan depends upon its central bank (BOJ) Bank of Japan, to sustain its economy. Every year on an average, the government of Japan invests 16.4% of the GDP into the developmental projects of the country. However, considering this from the economic point of view it is deliberate that such huge spending from tax revenue will slow down the economic growth. To boost the growth, BOJ endeavors to hold back a low interest rate. This is achieved by indulging in continuous buying of government debt, which as a strategy is closer to the "Quantitative Easing" policy followed by the Federal Reserve, USA.

Presently, out of 9 trillion government bonds, BOJ possesses 50% of the issued bonds. The discount rate of the bank is 0.3% and it has pledged to keep rates low. As a result of this, people anticipate low rates and the prices to go further down, which strengthens the reason for the deflationary environment. The anticipation indicates that as the price rises every time, consumers spending stops. People expect that prices will drop further. Due to this, businesses are not in a position to either increase the prices or employ new workforce. Employees have no possibilities to get increments, and as a result they keep saving, whatever little bit that they could.

BOJ desires to keep the Yen value at a low. However, whenever the value of Yen decreases, it inversely impacts the value of import goods and thus leads to inflation. In 2014 the currency value of USD increased by 15%, but the increase never reflected in the value of imports from the USA.

Adding fuel to the fire, downward spiral of oil prices has kept the prices to a low, which has made the deflation more worse. BOJ and the Japanese government have been making an effort to invigorate the growth through different measures as monetary policy and expansionary fiscal policy. Since it is a String of Pearls, the outcome has been that Japan has landed into the classic Liquidity Trap.

Japan, during the 1990s, encountered a monetary disaster when they went through the breaking of an illusion. The root of this predicament ought to have been strewn in the time period wherein the commercial non-interference that was introduced in the 1980s prior to the emergence of asset fantasy.

18.3 Features of Japanese Economy

Given below are the various components that impede Japan's economic stride. Leadership and the economic experts need to resolve the issues to bring back the growth path.

Keiretsu Strategy An organized interlinked relationship that prevails between the producers, vendors, dealers, and distributors. It offers the producers control and power akin to Monopoly that allows them to hold the supply chain in order. It lowers the influence of free market players. It is almost impossible for new entrants, innovators to engage with the Keiretsu, which is a low-cost initiative. Keiretsu casts down FDI. The advantages that Keiretsu provides a Japanese company make it almost impossible for non-Japanese companies to even think of competing with a Japanese organization.

Confirmed Lifelong Employment Japanese companies hire university and college graduates, who stay with them till superannuation. This can be understood as the job retention of employees till their retirement age by the same company. On an average throughout 25,000,000 employees are benefitted through the system. The workforce possesses obsolete skills, just cling onto the job till they reach retirement. This severely hampers the competitiveness of the corporates, as well as profitability since their salaries are increased artificially. However, 2008 recession severely dented the profitability and this culture has changed post-2008 recession. Commencing 2014, though only 8.8% of the Japanese corporations continuing with this policy, however its influence still exists.

Aging Population It is an indicator of low demand which cannot help to steer growth. Compared to the younger population, older population does not buy cars, houses, and consumer products. Government is burdened with the compulsion to pay more towards superannuation privileges than it collects through income tax from active employees. This means that the expenses in the form of retirement benefits will exceed the income earned through income taxes from the employed population. Added to this, dwindling population does not help the cause. Compared to the current population in 2015, Japanese population will shrink by 30% in 2065. An inflow of younger families would greatly help the economic growth. Japanese companies would be forced to depend on workforce that would be available in South Asian countries, who would be willing to work in Japan as expatriates. However, their earnings would be sent back to their parent country, which would again hamper Japan's progress.

Cheap interest rates are a result of (BOJ)'s **Yen carry trade**. A rational investor always thinks of taking the currency trade in Yen which benefits at a lower cost and trades the same in the form of British Pound and USD where they can reap better profits in return. The steadfastness of the Japanese currency maintains its value always complex than prediction of the central bank which results in deflation and minimized exports.

Growing Debt to GDP Ratio Which infers that Japan will be in debt twice in proportion to what it produces annually**.** As a consequence of this, BOJ is the massive owner of the debt. BOJ's continuing bond-buying policies have allowed Japan as a nation to indulge in spending without having a concern about the higher interest rates that are challenged by fidgety lenders. Regrettably, Japanese government's payout has not contributed in a significant improvement to the economy.

In both the financial years of 2015 and 2017, Japan also holds the dubious distinction, albeit briefly as the greatest holder of USA debt. This is purposeful action to keep the value of Yen under control and to promote the volume of exports in USD.

Globally Japan is one of the biggest total importers of food, despite having around one-third of their land area being fertile, productive, and cultivable. They are also net importers of energy products. BOJ's monetary policy rate at present remains at −0.1%. Some leading nation's central banks, such as the Swiss National Bank, do maintain lower policy rates, however, an absence of profitability for bank has restricted the BOJ to reduce the rates further down. The policy review puts forward by the BOJ with an intent to stem the downslide of an undesirable negative interest rate policy, nevertheless the results have been relatively little.

Yet, Japan's deflation might in the due course reverse on its own as well. As stated by BOJ's governor, Haruhiko Kuroda, deflation has been due to temporary reasons. Though Japan is probably expected to get out of the deflation spiral, the BOJ does not expect to attain its inflation target of 2% prior to 2024.

18.4 USD/JPY Currency

The USD/JPY pair is one of the most popularly traded currency pairs in the world, due to the popularity of US dollar throughout the world and JPY throughout Asia. It is referred to as "The Gopher." It possesses a high liquidity, indicating that traders can buy and sell the currency pair in huge volumes without the price fluctuating too much in its exchange rate. It also boasts one of the compacted spreads in the forex market, lowering the total costs of the trade (Fig. 18.1).

Since the time period from May 2020 to May 2021 this pair has traded with the conversion rate of one USD to Japanese Yen between 103.00 and 111.000 (Fig. 18.2).

The USD/JPY pair being one of the most closely traded currency pair in the world. In the last 5 years as could be inferred from the above, this pair has traded in sideways market. Referred as a *sideways market*, or *sideways* drift, this scenario

Fig. 18.1 Currency movement: May 2020 to May 2021. Source: https://tradingeconomics.com/japan/currency

Fig. 18.2 Currency Movement May 2015 to May 2021. Source: https://tradingeconomics.com/japan/currency

Fig. 18.3 Japanese currency index (measure against basket of six currency namely USD, GBP, EUR, CAD, CHF, AUD). Source: https://tradingeconomics.com/japan/currency

occurs where the price of a security *trades* within a fairly stable range without establishing any well-defined shifts over some considerable time period. Price action instead sways in a horizontal range or channel, with neither the bulls nor bears taking command of prices. The resultant meaning is that the big hedge funds and central bank did not involve during this period. In Jan 2021, the Japanese central bank decided to intervene in the market. As could be observed and inferred from the chart, Japanese Yen became weaker after the (BOJ) central bank intervention since Jan 2021 (Fig. 18.3).

18.5 Conclusion

Bank of Japan is one central bank, which possesses the experience of increasing its inflation target. In January 2013, BOJ (Bank of Japan) had increased its inflation target to 2% from 1% with the aim to put an end to persistent disinflation which has existed for more than a decade. Other central banks that are interested in inspecting the scenario of raising their inflation target in future could gain valuable lessons from the BOJ experience.

BOJ's supposition around inflation calls attention to the opposition that Japan's economy is confronted with. The pandemic will keep the output low as most of the buyers remain justifiably careful and attentive, as the risk of infection remains almost high. Still, uncommon incentive schemes in Japan and abroad are anticipated to augment GDP growth in the second and third quarter of the year. Systemic refinements may contribute little for growth this year but they do possess the ability to improve performance and the balance of trade in the long drawn out battle.

Given the prevailing scenario, the natural query that comes to a reader's mind is

1. Can this pair (USD/JPY) be traded in the long term?
2. Can this pair (USD/JPY) provide investors with the rate of return that could be similar to the other most traded currency pair?
3. What are the benefits of trading in a sideways market?

References

1. M. McFarlin, Top forex fundamentals. Futures: News, analysis & strategies for futures. Opt. Deriv. Traders **40**(4), 46–47 (2011)
2. J.V. Rosenberg, L.G. Traub, Price discovery in the foreign currency futures and spot market. J. Deriv. **17**(2), 7–25 (2009)
3. S. Schulmeister, Components of the profitability of technical currency trading. Appl. Financ. Econ. **18**(11), 917–930 (2008). https://doi.org/10.1080/09603100701335416
4. K. Pukthuanthong, D. Thiengtham, T. Walker, *Why Venture Capital Markets Are Well Developed in some Countries but Comparatively Small in Others: Evidence from Europe* (2007), pp. 33–49, ISBN 9780750682596. https://doi.org/10.1016/B978-075068259-6.50007-5
5. S. Nison, *Japanese Candlestick Charting Techniques: A Contemporary Guide to the Ancient Investment Techniques of the Far East* (Penguin, 2001)
6. A. Northcott, The complete guide to using candlestick charting: How to earn high rates of return—Safely. Atlantic Publishing Company 2009 (2009). https://doi.org/10.3905/JOD.2009.17.2.007
7. G. Cassel, *Money and Foreign Exchange After 1914* (Constable, London, 1922) Creative Media Partners, LLC, 2015, ISBN 1297383133

Chapter 19
Broca's Area of Brain to Analyze the Language Impairment Problem and Behavior Analysis of Autism

Md Ashiqul Islam, Rafat Karim, Faruq Ahmed, Maksuda, Md Sagar Hossen, and Shamima Akter

19.1 Introduction

The brain is a tremendous three-pound organ that controls all functions of the body, interprets data from the surface world. The human brain may be a hub for specific primary tasks. When our brain does not work properly or fails to complete his tasks our science contemplates it as an unfit brain. In our trendy science day by day we wish to understand a couple of human brain and behavior. During this continuation, we tend to see that the ordinary brain and unfit brain have some variations. The spectrum disorder syndrome people face some difficulties [1]. They cannot socially interact with people very well, the communication gap, and abnormal behaviors are facing in their faces.

In the frontal lobe, Broca's area is one of the reasons that play an important role in language production. Though its precise linguistic functions are still a bit unclear. It is named by physician Paul Broca. In this problem with aphasia reading and writing are also impaired but language comprehension is typically relatively preserved. Some symptoms are involved with producing movements like the tongue and mouth that help speech to be produced [2]. And also some other symptoms are that is involved producing grammar, verbal memory, syntax. Broca's area also has some linguistic and non-linguistic functions. It plays a role in language comprehension,

M. A. Islam (✉) · R. Karim · F. Ahmed · Maksuda · M. S. Hossen
Daffodil International University, Dhaka, Bangladesh
e-mail: ashiqul15-951@diu.edu.bd; rafat15-867@diu.edu.bd; faruq15-865@diu.edu.bd; maksuda15-854@diu.edu.bd; sagar15-1504@diu.edu.bd

S. Akter
George Mason University, Manassas, VA, USA
e-mail: sakter5@gmu.edu

P. M. Jeyanthi et al. (eds.), *Decision Intelligence Analytics and the Implementation of Strategic Business Management*, EAI/Springer Innovations in Communication and Computing, https://doi.org/10.1007/978-3-030-82763-2_19

movement, and even understanding the movement or actions of others and the overall function seems to be more complex.

One of the important regions is known as Wernicke's area. It is situated in the cerebral cortex. And this has been considered important to language comprehension and it produces meaningful speech [3].

Autism now becomes a great problem in the running condition of the world. But if it occurs for the brain and the particular portion named Broca's area, then it causes some important issues. Already we have discussed Broca's area and what it affects language production. So our main focus is to identify the problem. For that reason, we have to analyze the problem with this kind of patient and understand the issues. Besides that, we have to find the issue that occurred with the behavior analysis of autistic people.

So, we designed a methodology to identify the impairment problem in Broca's area and behavior analysis of autistic people. In our design, there will be some functionalities that may reduce the perspective problem with real-life patients attitude what we have analyzed from the dataset. In this paper, we are also given some functionality on how to solve this problem and what can we do to reduce the spectrum disorder problem.

In our dataset, there is a total of 20 attributes with the class attribute. Depending on the class attribute we can take the final decision. And there is a total of 1188 instances in our dataset. Depending on all of 19 attributes our algorithms which we have used for those will produces results. Now for our dataset, we have used some algorithms. And these algorithms are known as machine learning algorithms. We have used Decision Stump, Naïve Bayes, REP Tree, Logistic Classifier, Classification Via Regression, Decision Table, Random Forest, AdaBoostM1 algorithm. These algorithms are known as machine learning algorithms. Some of these algorithms are the best algorithm for classification. Because for obtaining the final output of our work we need to classify our dataset. And some of the algorithms are known as probabilistic algorithm classifiers. Depending on the probability of statistics those predict the result. So by using these algorithms we have got some terms those are accuracy, precision, recall, and F-Measure. These parts will be in our result section. Depending on the accuracy we can say that for our dataset one particular algorithm produces good accuracy that means that the algorithm is best for our dataset. So in our paper, we have discussed all of the parts depending on the results we have got from these algorithms and then some functionality on how to solve this problem.

19.2 Related Work

Farjana Khanum [3] analyzed Apache's symptoms to investigate how language and cognition behavior affects it. Get interviewed three patients to prove that the brain's production of a trauma disrupted normal language production.

Alfredo Ardila [4] analyzed the connection of language-related BA47 (pars orbitalis). Prove BA47 (pars orbitalis) language plays a central role in the production system. The results suggest BA47 participates in an extended Broca's system.

J.M.S. Pearce [2] said maximum of the papers in Bokor's work mentions the brief of his 1861 paper references.

David Embick [1] identified specific cortical regions using an experimental MRI method. For experimental design contrasted sentences containing grammatical errors with sentences containing spelling errors, ungrammatical sentences produced more activation in cortical language areas. Nowadays direct evidence of synthetic specialization is provided for broker area and the existence of individual modules establishes the knowledge of our language.

G. C. Imaezue [5] analyzed the role of broker territory is very important for linguistic and non-linguistic processing. Second, it has discussed the role of Broca's region. Third, a novel hypothesis is proposed for his experiment.

Patrik Fazio [6] studied and gained knowledge that the front aphasics are not affected by apraxia and it has capable of accurately encoding the regard human activity.

Alfredo Ardila [7] especially highlighted that there are two home systems and also its two separate levels of language. And this language is supported by the basis of two different teaching strategies and on the formation of different brain structures.

W.B.Groen [8] aims to review the available data on the structural work to provide a more integrated view of the language phenotype of autism, as well as the linguistic phenotype and its underlying neural deficits.

Tager-Flusberg [9] focused on the use of grammatical shapes in phonological processing and discusses the current neuromedicines and genetic studies of autism.

G.J Harris [10] explores brain activation patterns of adult male autism sufferers and ASD groups significantly reduced broker activation but increased left temporal activation.

Charlotte Kuepper et al. [11] identified reduced subsets of five behavioral features for the whole sample which showed good specificity and sensitivity and reached performance.

Mohamed A. Saleh [12] presented a model of the emotion-aware robot-assisted therapy, which is expected to ease the prediction and recognition of the emotion and behaviors of autistic children and enhances robot intervention during rehabilitation.

Dadang Eman et al. [13] provided a review on autism spectrum disorder by using a machine learning algorithm. Based on the results obtained, the most widely used algorithm support vector machine (SVM) get of 65.75% accuracy.

Thabtah [14] proposed a new machine learning method and collected three data sets related to children, adolescents, and adults.

Neda Abdelhamid et al. [15] employed oversampling techniques, such as Synthetic Minority Oversampling (SMOTE), and undersampling techniques Random Forest. Those are to measure the impact of these on the performance.

Azian Azamimi Abdullah et al. [16] selected three supervised machine learning algorithms, which are Random Forest, Logistic Regression, and K-Nearest Neighbors with K-fold cross-validation.

Said Baadel et al. [17] proposed a method that identifies potential autism cases and applied intelligent classification approaches such as Artificial Neural Network (ANN), Random Forest, Random Trees, and Rule Induction. These classifiers are useful as they are exploited by diagnosticians.

Fadi Thabtah et al. [18] analyzed 37 different ASD screening tools to identify possible areas and investigated different criteria associated with existing screening tools.

Noora Saleem Jumaa et al. [19] first collected patient's data and then send to the server. After that they receive data from the client and classify it with Naive Bayes algorithm if the patients has ASD or not.

Ngahzaifa Ab Ghani et al. [20] proposed a model using fuzzy agent. The fuzzy inputs five categories, which are Communication, Gross Motor, Fine Motor, Problem Solving, and Personal Social.

Md. Ashiqul Islam et al. [21] analyzed the risk factor of chronic kidney disease uning machine learning approaches. They found hemoglobin as the highest risk for the chronic kidney disease.

Md. Sagar Hossen et al. [22] recognized the papaya disease using convolutional neural network. They are applying VGG-16 technique to predict the disease. It can help them to cultivate more fruits.

Md. Ashiqul Islam et al. [23], they applying convolutional neural network based approach to predict the paddy leaf disease. VGG-19 is apply to recognize the disease to cultivate more disease free crops and increase their GDP.

Md. Ashiqul Islam et al. [24] applied many machine learning technique to analyze and classify the papaya disease. They are working on five major papaya disease to recognize the disease for better cultivation (Table 19.1).

19.3 Relation Between Language Processing and Brain

By understanding the speech and then processing the language our brain gets an acknowledgment. After getting the acknowledgement our brain system started working. That means there is a massive relation between brain and language processing [5].

In the brain, different regions exist and these regions are participating in different part for completing the language processing. One of the basic anatomies is structural anatomy. It has a connection with the brain. Language can handle the structure of the human brain, and it is known to many other brain linguistic abilities (Fig. 19.1).

In our brain there exist two parts that help to process language. One is Broca's area which helps for speech and another is Wernicke's area which helps for understanding language. Broca's area is responsible for language expression and when it damages people tend to have trouble to produce speech, this is called nonfluent aphasia or Broca's aphasia. Broca's aphasia means broken speech and aphasia is any type of disorder that disrupts language [3]. Conversely when Wernicke's area is temporally damaged, then it is called Wernicke's aphasia that means jumbled

Table 19.1 Research gap of some existing system

Paper information	Research gap
Kuepper et al. [11]	1. Results cannot be generalized to the entire ASD spectrum [11] 2. Their outcome criterion was not independent of the features utilized for building the prediction algorithm [11]
Thabtah et al. [14]	One of the limitations of this article is not including instances related to toddlers as these are rare and difficult to obtain [14]
Baadel et al. [17]	1. Used imbalanced dataset [17] 2. Datasets limited size
Abdelhamid et al. [15]	Does not compare the variations of SMOTE, to see what impact it has had on many different versions of the autism dataset [15]
Razali et al. [19]	Based on the proposed fuzzy agent-based model, an intelligent screening tool will be developed. One fully functional screening tool will be able to diagnose the level of severity and category [19]
Fadi Thabtah et al. [20]	1. Their applied machine learning in ASD research has not considered conceptual, implementation, evaluation, and data related issues [20] 2. Their training dataset imbalances
T. Akter et al. [25]	ASD data was militated. In the future, they will analyze more data to improve the detection of ASD [25]
Erkan et al. [26]	Number of data samples is not enough which is their study limitation [26]
Matsuzaki, Junkoet et al. [27]	In their study there was no clinical control group of children with intellectual disability [27]

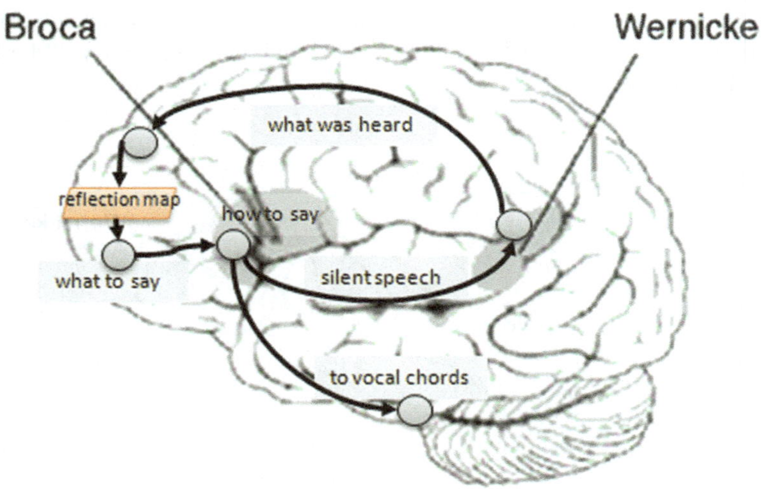

Fig. 19.1 Brain map of language processing [3]

speech, it is quite a different pattern of behavior with Broca's aphasia. When both of these aphasia are present, then it becomes global aphasia.

19.4 Functionality of Broca's Area

Broca's area is liable that is produced of speeches. It regulates the motor functions that are related to speech production. If the Broca's area of the brain is damaged people can not understand the word properly, they need to struggle to deliver their speech.

The ability to understand and speak the language is related to several areas of the cerebral cortex. Basically, auditory cortex can first perceive the spoken language and the visual cortex can process in written text, or sign language [7] (Fig. 19.2).

Both information is sent to the Wernicke's area, in the temporal lobe [2, 3] then it matches the vocabulary of the person in memory and converts the information to a signal, the signals are transmitted through a bundle of nerve fibers called the arcuate fasciculus, into the frontal lobe broker region [10]. Broca's area is liable that is produced of speeches. Output goes from broca's area to the motor cortex, that is, control of muscle movements for speech (Fig. 19.3).

A language disorder caused by brain damage is called aphasia. Sensitive or receptive lesions are caused by aphasia in the region of Warwick. They can speak at a smooth pace, but their speech is often involuntary. They can talk normally, but they cannot explain the sentence. The erosion of the Bokor region is usually related to the content created in the content vocabulary. For example, children with Broca's aphasia may say something like, "Eat, rice. Mom." meaning to say, "Mom, I'm hungry, I'll eat rice" [4, 7].

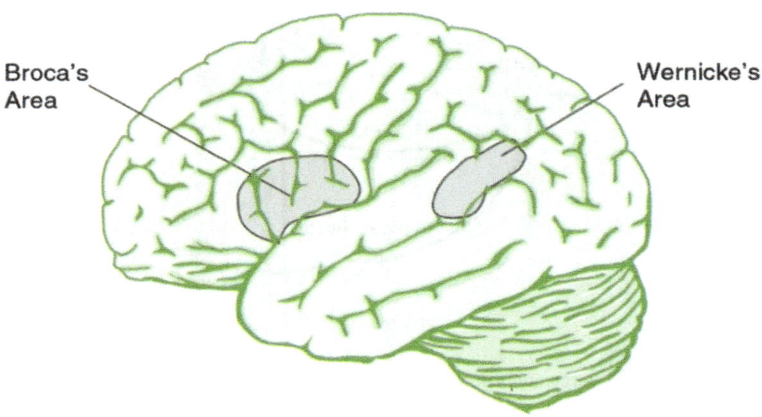

Fig. 19.2 Broca's area of brain [7]

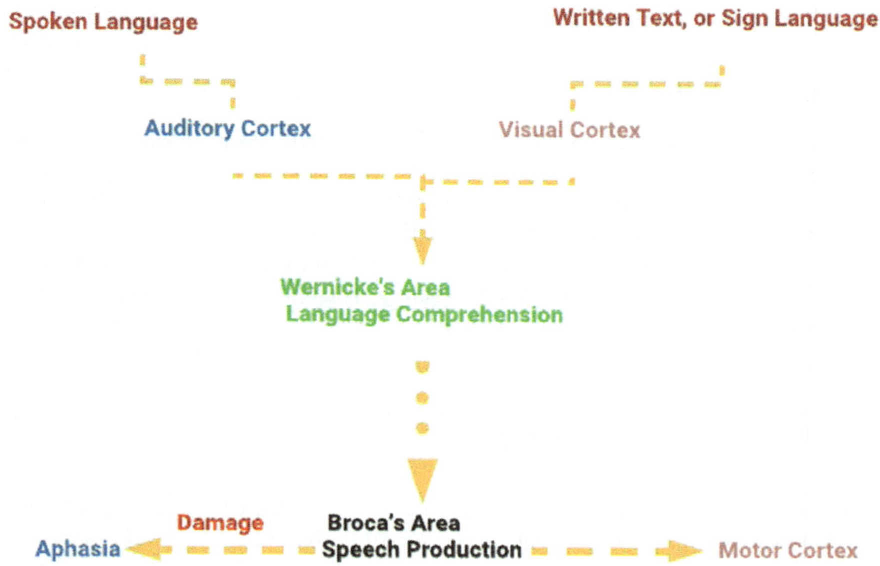

Fig. 19.3 Broca's area of brain functionality [11]

19.5 Language Impairment Problem Solution

Autism spectrum disorder is a brain development condition, it manifests itself as difficulty in communicating and socializing with others [3, 6, 9]. Autism is a spectrum disorder which affects an individual through the patient's life. There are many symptoms of autism as like avoidance of eye contact, ignore caregivers when called by name, facial expression, body posture, the lake of talk, social interaction.

19.5.1 Medical Checkup

First, we have to take them near to a doctor. Otherwise, the doctor can diagnose their conditions if they have any kind of impairment. So it should be our first step.

19.5.2 Language Therapy

We can easily try to fix this problem with language therapy. But in this case, an important parameter will be children's age. We have to early intervention for a tremendous outcome.

19.5.3 Home Care Options

Working with child at home can help. Here are some tips:

1. We have to speak clearly with children. And when will ask a question that should be slowly and concisely.
2. For their response, we have to keep our patients because they need time.
3. The situation should be relaxed in reducing their anxiety.
4. After giving a command we should ask the child for their instruction to put on.
5. With teachers, we have to contact frequently. The child will not easily cop up in the class easily.

19.5.4 Psychological Therapy

To understand and communicate can be frustrating and can trigger acting episodes. Counseling may be needed to address behavioral problems.

19.6 Methodology

As we described that our dataset is behavior analysis of autism data. Our target is to predict the accuracy depending on the attributes of our datasets. Understanding the datasets and finding the dependency of attributes among themselves, we have to do the following steps.

For our dataset, we have done data visualization, classification, boost, and at last regression analysis. For classification we have used Decision Stump Classifier, Logistic Regression Classifier, Naïve Base Classifier, REP Tree, Decision Table, RandomForest Classifier, AdaBoost, and Classification Via Regression (Fig. 19.4).

At first we input the data from the behavior analysis of autism dataset. Then we normalize the data and encoding the categorical data in numerical and preprocessing the dataset. After successfully preprocessing we split our data for training and testing validation. We are using 80% data for training and 20% data for testing. We are applying some classifier and regression model to analyze the data and classify the data for visualization. Then we can easily decide which patients have autism symptoms and which one has not. We can find out the decision easily by applying this method and find out the best fit model for our dataset.

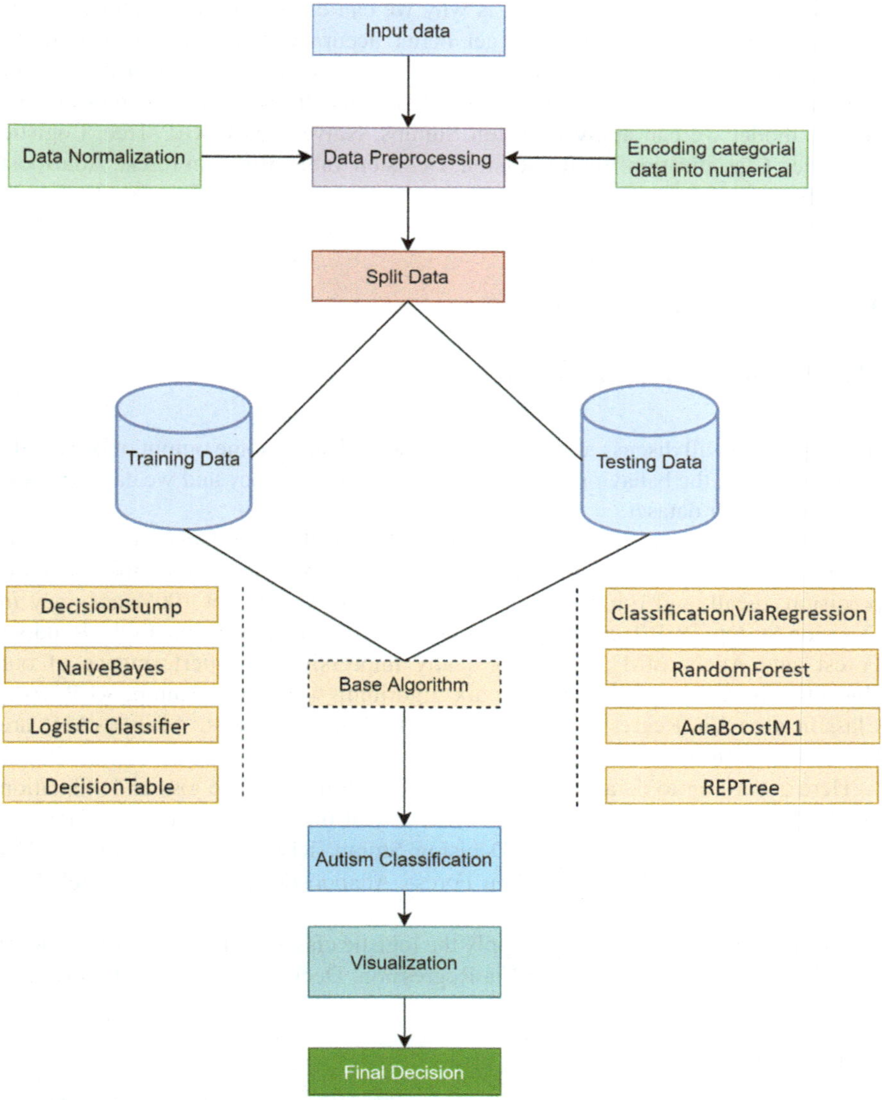

Fig. 19.4 Proposed model [16]

19.7 Dataset

We can collect our dataset from a GitHub repository. The location of GitHub repository is https://github.com/sagarcse1504/Behavior-Analysis-of-Autism. One important thing is that we have preprocessed our dataset and there have not any missing values or noisy values that is why we can easily gain better performance from our applying algorithm and get better accuracy and this can best fit for our dataset. In our dataset there are 1188 samples data and 19 attributes. We can analyze the behavior of this dataset to predict the accuracy. To find out the best fit model we can apply Decision Stumps, Naive Bayes, REP Tree, Logistic Classifier, Classification Via Regression, Decision Table, Random forest, AdaBoost algorithms.

19.8 Results and Discussion

19.8.1 Behavior Analysis Result

In this part we will discuss about our result and will show some output of our result. We can analyze the behavior of autism to predict the accuracy and we can gain best fit model for our dataset.

From the experimental analysis we can observe that the dataset is preprocess and the accuracy of our algorithms is really good. We can analyze the accuracy, precision, recall, and F-measure in Table 19.2 and we can get 100% accuracy in Decision Stump, REP Tree, Classification Via Regression, Decision Table, Random Forest, and AdaboostM1 which are really impressive. The performance of our algorithm is really good and these six algorithms (Decision Stump, REP Tree, Classification Via Regression, Decision Table, Random Forest, AdaboostM1) are best fit for our model.

Here according to dataset and best model we found out the loss and validation graph. From Figs. 19.5, 19.6 we can observe that the loss and validation are less and we can get best accuracy in Decision Stump, REP Tree, Classification Via Regression, Decision Table, Random Forest, AdaboostM1, and these models best fit for our dataset.

By this training dataset we can apply the logistic classifier, Decision Stump, REP Tree, Naive Bayes, Classification Via Regression, Decision Table, Random Forest,

Table 19.2 Accuracy of the algorithm

Algorithm	Accuracy	Precision	Recall	F-measure
Decision Stump	100%	1.00	1.00	1.00
Naïve Bayes	98.4%	0.985	0.984	0.984
REP Tree	100%	1.00	1.00	1.00
Logistic Classifier	99.8316%	0.998	0.998	0.998
Classification Via Regression	100%	1.00	1.00	1.00
Decision Table	100%	1.00	1.00	1.00
Random Forest	100%	1.00	1.00	1.00
AdaBoostM1	100%	1.00	1.00	1.00

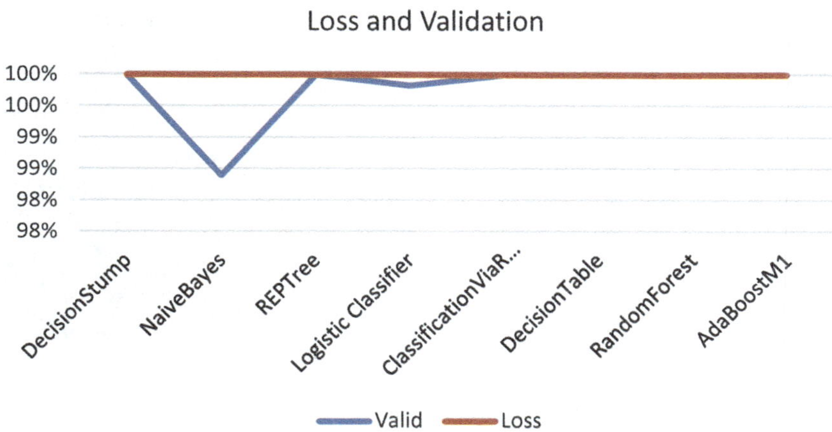

Fig. 19.5 Loss and validation graph

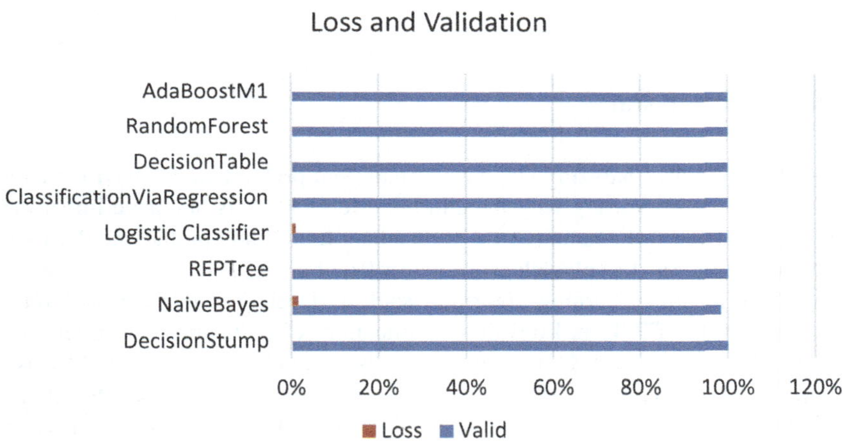

Fig. 19.6 Training and validation loss graph

AdaboostM1 and find some output of training accuracy, training validation, training loss, and testing validation [13]. Overall we can say that Decision Stump, REP Tree, Classification Via Regression, Decision Table, Random Forest, AdaboostM1 are best for behavior analysis of autism dataset and its accuracy is good 100%. It can be fit for this model. At last rerunning the model with using our classifier algorithm we found the accuracy from our model. There are lots of part in our codes, but we have highlighted some of our output here.

In Fig. 19.7 we can analyze the Rh factor behavior analysis of autism and we can observe that the percentage of Rh factor A (positive, negative) is high 43%, Rh factor B (positive, negative) is 26%, and Rh factor of C (positive, Negative) is 31%.

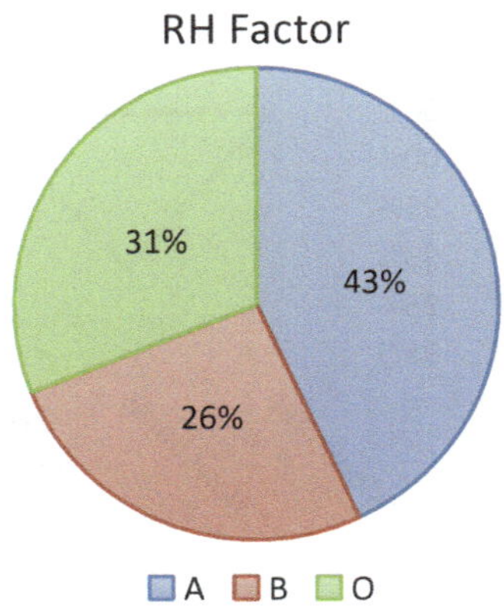

Fig. 19.7 Rh factor behavior analysis of autism

So, we can say that the Rh factor A (positive, negative) is very much possible to be affected in autism.

19.9 Comparison

In this paper we are analyzing language impairment problem and behavior analysis of autism. We are focusing why the children are facing this language impairment problem and what is the main reason of this disorder to find out that and discussing about some criteria to solve this problem [19]. For behavior analyzing we are using many machine learning algorithm such as Decision Stump, Naive Bayes, Random Forest, REP Tree, Logistic Classification, Classification Via Regression, Decision Table, AdaboostM1 algorithm and it is really good to known that Decision Stump, REP Tree, Classification Via Regression, Decision Table, Random Forest, AdaboostM1 perform very well and its accuracy is 100%. This algorithm performs well than other and it is the best model to fit in the behavior analysis of autism dataset and its training accuracy and loss and validation are also analyzed for its best fit.

19.10 Conclusion

Autism is a mysterious brain disorder among numerous abnormalities in brain function even after extensive injuries [28]. Language skills return to normal or near spontaneously during periods of weeks or months when the brain recovers from the physical assault [5, 6]. Speech therapy and counseling can be very effective treatment for continuous language problems. So, we suggest that broker territory is likely be the center of the brain network encoding classification regardless of their use of the structure action, language, or music. For this, we have to analyze the behavior of autism. At last, the patient's history and research findings show that it depends on the patient's recovery on their age and the nature of injuries.

References

1. D. Embick, *A syntactic specialization for Broca's area* (Massachusetts Institute of Technology, Cambridge, 2000)
2. J.M.S. Pearce, Broca's aphasiacs. Eur. Neurol. **61**, 183–189 (2009). https://doi.org/10.1159/000189272
3. Influence of Broca's aphasia and Wernick's aphasia on language disorder. Double Blind Peer Rev. Int. Res. J. **14**(9), (2014)
4. A. Ardila, B. Bernal, M. Rosselli, Should Broca's area include Brodmann area 47? Psicothema **29**(1), 73–77 (2017)
5. G.C. Imaezue, Brain localization and the integrated systems hypothesis: evidence from Broca's region. J. Behavior. Brain Sci. **7**(11), 511–519 (2017). https://doi.org/10.4236/jbbs.2017.711036
6. P. Fazio, A. Cantagallo, L. Craighero, A. D'Ausilio, A.C. Roy, T. Pozzo, F. Calzolari, E. Granieri, L. Fadiga, Encoding of human action in Broca's area. Brain **132**(7), 1980–1988 (2009). https://doi.org/10.1093/brain/awp118
7. A. Ardila, There are two different language systems in the brain. J. Behavior. Brain Sci. **1**, 23–36 (2011). https://doi.org/10.4236/jbbs.2011.12005
8. W.B. Groen, M.P. Zwiers, R.J. van der Gaag, J.K. Buitelaar, The phenotype and neural correlates of language in autism: An integrative review. Neurosci. Biobehav. Rev. **32**(8), 1416–1425 (2008)
9. H. Tager-Flusberg, Defining language phenotypes in autism. Clin. Neurosci. Res. **6**(3–4), 219–224 (2006)
10. G.J. Harris, C.F. Chabris, J. Clark, T. Urban, I. Aharon, S. Steele, L. McGrath, K. Condouris, H. Tager-Flusberg, Brain activation during semantic processing in autism spectrum disorders via functional magnetic resonance imaging. Brain Cogn. **61**(1), 54–68 (2006)
11. C. Kuepper, S. Stroth, N. Wolff, F. Hauck, N. Kliewer, T. Schad-Hansjosten, I. Kamp-Becker, L. Poustka, V. Roessner, K. Schultebraucks, S. Roepke, Identifying predictive features of autism spectrum disorders in a clinical sample of adolescents and adults using machine learning. Sci. Rep. **2020**, 10 (2020). https://doi.org/10.1038/s41598-020-61607-w
12. Int. J. Adv. Trends Comp. Sci. Eng. **8**(1), 6 (2019). http://www.warse.org/IJATCSE/static/pdf/file/ijatcse6381.62019.pdfhttps://doi.org/10.30534/ijatcse/2019/6381.62019
13. D. Eman, A. Emanuel, *Machine learning classifiers for autism spectrum disorder: a review* (Information Systems and Electrical Engineering (ICITISEE), Yogyakarta, 2019). https://doi.org/10.1109/ICITISEE48480.2019.9003807
14. F. Thabtah, D. Peebles, A new machine learning model based on induction of rules for autism detection. Health Informatics J. **26**, 146045821882471 (2019). https://doi.org/10.1177/1460458218824711
15. N. Abdelhamid, A. Padmavathy, D. Peebles, F. Thabtah, D. Goulder-Horobin, Data imbalance in autism pre-diagnosis classification systems: an experimental study. J. Inf. Knowl. Manag. **19**, 2040014 (2020). https://doi.org/10.1142/S0219649220400146
16. A. Abdullah, S. Rijal, S. Dash, Evaluation on machine learning algorithms for classification of autism spectrum disorder (ASD). J. Phys. Conf. Ser. **1372**, 012052 (2019). https://doi.org/10.1088/1742-6596/1372/1/012052
17. S. Baadel, F. Thabtah, J. Lu, A clustering approach for autistic trait classification. Inform. Health Soc. Care **45**, 309 (2020). https://doi.org/10.1080/17538157.2019.1687482

18. F. Thabtah, D. Peebles, Early autism screening: a comprehensive review. Int. J. Environ. Res. Public Health **16**, 3502 (2019). https://doi.org/10.3390/ijerph16183502
19. N. Razali, N. Ab Ghani, S. Hisham, Intelligent autism screening using fuzzy agent. Lecture Notes Electr. Eng. **632**, 495–503 (2020). https://doi.org/10.1007/978-981-15-2317-5_42
20. F. Thabtah, Machine learning in autistic spectrum disorder behavioral research: A review and ways forward. Inform. Health Soc. Care **44**(3), 278–297 (2019). https://doi.org/10.1080/17538157.2017.1399132
21. M.A. Islam, S. Akter, M.S. Hossen, S.A. Keya, S.A. Tisha, S. Hossain, Risk factor prediction of chronic kidney disease based on machine learning algorithms, 2020 3rd International Conference on Intelligent Sustainable Systems (ICISS), Thoothukudi, India, 2020, pp. 952–957, doi: https://doi.org/10.1109/ICISS49785.2020.9315878
22. M.S. Hossen, I. Haque, M.S. Islam, M.T. Ahmed, M.J. Nime, M.A. Islam, Deep learning based classification of papaya disease recognition, 2020 3rd International Conference on Intelligent Sustainable Systems (ICISS), Thoothukudi, India, 2020, pp. 945–951, doi: https://doi.org/10.1109/ICISS49785.2020.9316106
23. M.A. Islam, M.N.R. Shuvo, M. Shamsojjaman, S. Hasan, M.S. Hossain, T. Khatun, An automated convolutional neural network based approach for paddy leaf disease detection. Int. J. Adv. Comp. Sci. Appl. **12**, 1 (2021). https://doi.org/10.14569/IJACSA.2021.0120134
24. M.A. Islam, M.S. Islam, M.S. Hossen, M.U. Emon, M.S. Keya and A. Habib, Machine learning based image classification of papaya disease recognition, 2020 4th International Conference on Electronics, Communication and Aerospace Technology (ICECA), Coimbatore, India, 2020, pp. 1353–1360, doi: https://doi.org/10.1109/ICECA49313.2020.9297570
25. T. Akter et al., Machine learning-based models for early stage detection of autism spectrum disorders. IEEE Access. **7**, 166509–166527 (2019). https://doi.org/10.1109/ACCESS.2019.2952609
26. U. Erkan, D. Thanh, Autism spectrum disorder detection with machine learning methods. Curr. Psychiatr. Rev. **15**, 297–308 (2019). https://doi.org/10.2174/2666082215666191111121115
27. J. Matsuzaki, E.S. Kuschner, L. Blaskey, L. Bloy, M. Kim, K. Matthew, J.C. Edgar, D. Embick, T.P.L. Roberts, Abnormal auditory mismatch fields are associated with communication impairment in both verbal and minimally verbal/nonverbal children who have autism spectrum disorder. Autism Res. **12**(8), 1225–1235 (2019)
28. N. Jumaa, A. Salman, D.R. Al-Hamdani, The autism spectrum disorder diagnosis based on machine learning techniques. J. Xi'an Univ. Architect. Technol. **12**, 575–583 (2020)

Index

P. M. Jeyanthi et al. (eds.), *Decision Intelligence Analytics and the Implementation of Strategic Business Management*, EAI/Springer Innovations in Communication and Computing, https://doi.org/10.1007/978-3-030-82763-2

The manufacturer's authorised representative in the EU is Springer Nature Customer Service Centre GmbH, Europaplatz 3, 69115 Heidelberg, Germany. If you have any concerns regarding our products, please contact ProductSafety@springernature.com

Printed and bound by CPI Group (UK) Ltd, Croydon, CR0 4YY
28/09/2026
02224090-0007